Social Work and Prostitution

Martin Albert · Julia Wege
Editors

Social Work and Prostitution

Professional Approaches in Theory and Practice

Springer

Editors
Martin Albert
SRH Hochschule Heidelberg
Heidelberg, Germany

Julia Wege
RWU Hochschule Ravensburg-Weingarten
Weingarten, Germany

ISBN 978-3-658-37760-1 ISBN 978-3-658-37761-8 (eBook)
https://doi.org/10.1007/978-3-658-37761-8

The translation was done with the help of artificial intelligence (machine translation by the service DeepL.com). A subsequent human revision was done primarily in terms of content.

This Springer imprint is published by the registered company Springer Fachmedien Wiesbaden GmbH, part of Springer Nature.
The registered company address is: Abraham-Lincoln-Str. 46, 65189 Wiesbaden, Germany

Contents

List of Figures

Introduction

Martin Albert and Julia Wege

Prostitution is an extremely complex social field which, for many reasons, defies clear explanatory patterns. The underlying contradictions in dealing with the topic already lie in the act of prostitution itself. It touches on central questions of sexuality, intimacy, ethical values and individual attitudes in the context of gender relations. The entire field appears confusing and is highly tabooed, so that a comprehensive analysis is hardly possible. In this respect, it is also only possible with limitations to systematically present the diversity of social work associated with it. This is surprising in so far as there is a long historical connection and probably no other discipline is more closely active and present in the field of female prostitution than social work. The focus of social work efforts has always been on women and the struggle for their rights. However, just as there is little well-founded and verified knowledge about the actual lives of women, there is almost no view of how professional social workers actually fare in this professional field. Which framework conditions they are exposed to, which methodological concepts are important depending on the target group, and which professional attitudes fundamentally flow into the work, have hardly been named concretely in the professional discourse so far. Just as the field of prostitution or sex work is socially marginalized, this is also reflected in the field of publicly perceived social work. In

M. Albert (✉)
Heidelberg, Germany
e-mail: martin.albert@srh.de

J. Wege
Mannheim, Germany
e-mail: julia.wege@rwu.de

M. Albert, J. Wege (eds.), *Social Work and Prostitution*,
https://doi.org/10.1007/978-3-658-37761-8_1

a certain respect, one can speak of a partial and publicly effective "invisibility" on this basis, which affects both social work and women.

At present, the topic is controversially discussed under the social modernization processes and is thus subject to its own dynamics. The increased media interest often takes on voyeuristic features, which even professional social work cannot escape. This is accompanied by an increased pressure to justify why social work is involved in women's issues in the first place and what objectives are associated with this. The enlargement of the EU and the opening of borders has exacerbated the problems of poverty and immigration and has an impact on the prostitution sector in many different ways. The fact that the majority of women working in this field are migrants poses new methodological and professional challenges for social work and forces many counselling centres to expand their traditional counselling approaches. Like all areas of society, prostitution is also subject to the complex aspects of economisation, globalisation and the influence of the new media. Social work has to face these challenges in many ways and makes it interesting for society and politics as a profession to solve complex problems. Social work as a profession has developed enormously in recent years in terms of a methodological foundation and independent scientific character. Although these developments are continuously incorporated into the field of prostitution, they are hardly recognizable in the public and professional discourse. Basically, it is a narrowly defined field of work that has reached a high level of independence and professionalism, but neither in public nor in social work education is it given the necessary status.

On this basis, this book presents a first stocktaking in the context of social work and prostitution, focusing on professional social work in theoretical and practical terms. It was a central concern that the authors, who all have a high degree of theoretical and practical knowledge, present their view of the complex field based on social work methods and counselling approaches. This was done in relation to the underlying lifeworlds and needs of the associated target groups. The focus of the book is clearly on professional social work with women involved in prostitution. Both to distinguish and clarify the two areas, but also as an important complement to the methodological focus, we understand the two contributions that address social work with victims of trafficking for sexual exploitation. This seems to make sense, especially since there is a great need for clarification and information even within professional social work. All contributions have in common that they refer to case studies from practice and to basic social work methods such as individual case assistance, case management, group work, network work, community work and supervision. This has succeeded in placing the focus of professional action at the centre.

In the article "Social Work in the Field of Prostitution: Structural Development Trends in the Context of Organization, Social Space and Professional Role" **Martin Albert** points out the problem that only little well-founded information is available on prostitution. In addition, there is a lack of meaningful research results, publications, theoretical analyses and socio-spatial studies on the part of social work. Albert attempts to examine the counselling landscape in terms of organisation, sponsorship and funding in the context of developments over the past decades. It is critically shown that the existing organisational structures have historically grown in different ways and thus have a considerable influence on professional behaviour in everyday counselling practice. Furthermore, prostitution in the social space is analysed, also under the aspect of how far the constantly changing areas and milieus have an influence on the practice of social work. Based on different attitudes towards prostitution, specific role patterns of social work are described.

In her article "Prostitution in the Neighborhood: Development Concepts for a Design in Public Space", **Christiane Howe** refers to the different uses of a densely populated area and the role that the topic of prostitution plays here, based on the neighborhood of Berlin-Schöneberg. In a study carried out, residents, actors from different areas, prostitutes and clients of the neighbourhood were interviewed and asked about conflicts and solution strategies. The results were compiled and subsequently a citizens' exhibition was developed together with artists. One of the central questions of the investigation was how a constructive coexistence can look like in a tense neighbourhood with a high prostitution problem. First of all, the terms prostitution, human trafficking and neighbourhood are explained. The different lines of conflict in the prostitution scene are analysed and the role of social work is shown. Howe pleads for political negotiation processes and a stronger presence on the part of social work professionals.

The article "Drug-Using Sex Workers Are Service Providers: A Change of Perspective in Social Work in the Fight Against Sexualised Violence and Exploitation in "Drug Prostitution" " by **Kathrin Schrader** shows the difficult situation of drug-using sex workers. First, studies on the situation of violence of women in prostitution who are addicted to drugs are presented. Research results confirm that the affected clientele is exposed to an increased level of violence. The author differentiates here between physical violence, sexual violence and transgressions against oneself. The life situation of drug-using sex workers is described in detail and that social stigmatisation, experiences of violence by clients, low wages, the effects of drug addiction and a lack of social recognition all contribute to the fact that the women find themselves in a difficult milieu. Whether sex work is to be understood as work and what understanding women have of this is discussed in conclusion and supplemented with its own definition of drug-using sex workers.

The article "Social Work in the Context of the World of Prostitution: Professional Approaches in the Field of Tension of Different Systems and Actors" by **Julia Wege** shows which demands, tasks and goals social work has and how these are concretely shaped in the professional field of prostitution. In particular, it shows the different systems in which social work operates and has to react to them. A distinction is made between the system of the public or society, the system of cooperation partners or the professional public, and the system of the milieu or the living environment of the women. On this theoretical basis, the necessary and diverse competences of social work become apparent. The action concept of lifeworld orientation according to Thiersch should always be included in everyday counselling. Finally, questions for a reflected and cross-thematic individual case assistance are formulated and the future need for action is pointed out.

In their article "Outreach Social Work in the Field of Health Services for Sex Workers", **Elfriede Steffan** and **Tzvetina Arsova Netzelmann** describe the significant changes in the use of health services by sex workers that became apparent with the introduction of the Infection Protection Act (Infektionsschutzgesetzes – IfSG) in 2001. They provide a brief review of the development of the fields of activity of the health offices and specialist counselling centres, which have had a wide range of experience in the field of prostitution since the 1980s. Critically, however, it must be noted that STI and HIV counselling and medical services have been rapidly reduced as a result of the change in the law and that outreach work by the health offices is now only offered in a few cities. In this context, principles and criteria for successful outreach work are highlighted.

In the article by **Jacqueline Suter** and **Melanie Muñoz** "Sex Work and Social Work: A Instruction Manual", the field of action in which social work operates in Switzerland is shown. Starting from the political attitudes and legal regulations in Switzerland, the topic of sex work and migration is dealt with in particular. Using a case study of a client, it becomes clear what specific difficulties the women bring with them and what complex challenges social work is confronted with. The authors refer to the professionalism of social workers with the necessary qualifications, methods and specific forms of counselling. Social work operates in a difficult field of different conflicts of interest. Especially from a legal point of view, grey areas arise between the police, subsidy providers, politicians, prostitutes, clients and brothel operators.

In her contribution "Social Work with Prostitutes in Zurich: Illustrated by the Experiences of the Isla Victoria Counselling Centre", **Regula Rother** presents the prostitution scene in the political and social context of the city of Zurich and explains the fields of activity of the counselling centre. Based on this, the topics of power and powerlessness and the influence of organisational structures are discussed theoretically. In this context, social work acts on the basis of human rights. The mission statement of the counselling centre and the variety of methods in relation to individual case help, empowerment, group work, street work, social space orientation and community work are described in detail. Finally, challenges and basic conditions for Switzerland are discussed. Improved professional training, networking with other disciplines and an independent specialist unit that ensures political and professional development and cooperates concretely with actors from the universities could result in an immense boost to professionalism.

Julia Kempl addresses in her contribution "Career Entry in the Field of Female Prostitution: Challenges in the Context of Social Work Counselling" that the topic of prostitution is not part of the curriculum of universities for social work and thus female career entrants must first acquire a complex variety of methods and necessary basic knowledge. Furthermore, the access to the clientele via the method of street work is described and the challenges for the counselling specialist are discussed. A distinction is made between professional, methodological, personal and social competence, which must go hand in hand with a reflective and unbiased attitude. Due to immigration and the high proportion of migrants, intercultural competences are increasingly in demand. These competences can only develop through empathy and respect towards women. These competences as well as the combination of practical and theoretical knowledge are the optimal basis for an entry into the professional field of prostitution.

In their article "Social Work with Trafficked Persons for Sexual Exploitation in Germany", **Naile Tanis** and **Tabea Richter** show the current and legal political development in the protection of victims. An overview of the legal framework, especially in the European context, is given in detail. The specialised counselling centres organised in the KOK (Bundesweiter Koordinierungskreis gegen Menschenhandel e. V. – Federal Coordination Group against Human Trafficking) have developed and professionalised themselves in recent years by formulating common elementary goals as a basis for their work. A case study is used to illustrate the methodology of social work. The basis for the work with affected victims is the human rights approach. However, in order to be able to implement the work more professionally and to be able to help the victims in a far-reaching way, the requirements of the EU Directive on Preventing and Combating Trafficking in Human Beings and Protecting its Victims as well as the corresponding Council of Europe Convention on Action against Trafficking in Human Beings must be realised.

Katharina Kähler describes in her article "(Forced) Prostitution: Between Voluntariness and External Determination. Insights from the Perspective of Practical Social Work in a Specialised Counselling Centre Against Human Trafficking" the field of action with women who have become victims of human trafficking. First, the concept of voluntariness is examined under philosophical, socialisation-theoretical and legal aspects. Within social work, a discipline-specific and independent approach is of central importance. The tasks of an appropriate specialist counselling centre for victims of human trafficking include initial care, legal or financial clarification and, in addition, comprehensive psychosocial counselling and stabilisation. Case studies are used to illustrate the problems faced by victims and the role of social work. The corresponding milieu structures and dependencies pose great challenges for counsellors and social work. The development of core questions is intended to provide initial guidance in counselling practice. Of central importance are the competences of the counsellors, the demarcation from work, the comprehensive knowledge of structures in the prostitution milieu and the need to maintain a political independence.

In her contribution "Coaching, Supervision and Process Support of Counselling Centres for Sex Workers" **Angela Siebold** discusses the importance of external support in different contexts of social work in the field of prostitution. On the one hand, the role of the counsellor must always be reflected upon, and on the other hand, counselling centres are located in complex and often unsecured financial areas, so that external process support is of great importance. Siebold begins by clarifying the terms and distinguishes between coaching, supervision and process support and highlights central issues of supervision such as demarcation and professional distance. Equally important in this discourse appears coaching for leaders of counseling centers. While coaching is about the leadership role, content-related goals and tasks, process support for counselling centres addresses different processes within the contexts of society, institution and counselling system.

Julia Wege and **Martin Albert** present a scientific expertise entitled "Needs Analysis and Concept Development for a Counselling Centre in the Field of Prostitution for the City of Mannheim" in their separate contribution. Based on a master's thesis from 2010, corresponding scientific data on the prostitution problem in Mannheim was available. Based on this, further expert interviews were conducted and on-site analyses of other cities and projects were included in order to develop a tailor-made action concept on behalf of the Diakonisches Werk for the city of Mannheim. In doing so, the different fields of activity, target groups, general conditions and methods of social work are discussed and precisely pointed out. The close networking with political parties, police, municipal authorities and social institutions plays an essential role. The article concludes with concrete recommendations for the establishment of a counselling centre, which have meanwhile been implemented in the project “Amalie”.

It has been a particular objective of the authors’ contributions that the focus is on social work and its methodical practice of action. This required a high degree of openness and objectivity for the existing variety of topics. We would like to expressly thank all authors for their great commitment to the success of this book project. We would also like to thank Manuel Malcherowitz and Djana Herre, who supported us in the correction and formatting of the contributions.

Finally, we are pleased that, especially in the context of the complex and strongly tabooed field of action, this book describes for the first time a multitude of insights, empirical knowledge and competencies for professional social work and thus makes them accessible to the interested professional public. It is hoped that social work in the field of prostitution will continue to develop professionally and receive the necessary recognition and financial support from politics and society. Only on this basis can it make an important and sustainable contribution to ensuring that women are respected in their living environment and in their precarious social and legal position by society and the public, and that their dignity is made visible.

Martin Albert Social Worker (M.A.), is head of the social work program at the SRH Hochschule Heidelberg. His focus is on social work science, professionalization, migration and community work. He has conducted various scientific studies in the fields of street work, district development and prostitution.

Julia Wege Social Worker (M.A.), is head of the counselling centre “Amalie” – for women in prostitution in Mannheim. Since 2009 she has been a lecturer at the SRH Hochschule Heidelberg for individual case assistance, community work and trauma work. Her research and publications focus on prostitution, homelessness and community work.

Social Work in the Field of Prostitution: Structural Development Trends in the Context of Organisation, Social Space and Professional Role

Martin Albert

1 The State of Social Work in the Field of Prostitution

In the context of social work and prostitution, there is a lack of reliable data, comprehensive theoretical analyses, the evaluation of historical documents and well-founded socio-spatial or local studies. Even the actual number of women working in this field can only be estimated. The figure of 400,000 women in prostitution, which has been quoted since the 1980s, is repeatedly doubted. Kavemann and Steffan (2013) consider the figures to be inflated and assume 64,000–200,000 women. Somewhat more concrete are the figures from cities and municipalities, which primarily refer to police data. In the so-called "bright field", those women are registered who are in known brothels, running houses, date flats and on the street prostitution and pursue their trade. In the invisible "dark field" only vague estimates can be made. Prostitution of women with uncertain residence status, initiation via the internet or prostitution on illegal street prostitution hardly offer any clues about reliable findings. Prostitution is, however, perceptible in all territorial areas, i.e. also in rural and border regions. However, due to historical and legal developments, prostitution is more likely to be found in conurbations and urban areas.

M. Albert (✉)
Heidelberg, Germany
e-mail: martin.albert@srh.de

M. Albert, J. Wege (eds.), *Social Work and Prostitution*,
https://doi.org/10.1007/978-3-658-37761-8_2

The present contribution can only represent an attempt to systematically describe recognisable characteristics of the professional field. For this purpose, among other things, the aspect of the organization of the professional counselling centres and the associated financing and carrier structure will be presented in more detail on a macro level. On a meso level, the social or territorial space of prostitution in the context of social work will be considered. On a micro level, an attempt will be made to describe the influence of value attitudes on the professional role. The following basic theses can be formulated for the present contribution:

1. Prostitution is subject to a limited public perception and is largely characterised by a high degree of taboo and discrimination against women.
2. The field of prostitution is subject to an ongoing social discussion of values, the positioning of which can differ greatly depending on the viewpoint, professional discipline and moral concepts.
3. Women in the field of prostitution represent an extremely heterogeneous target group, which is particularly characterised by their different motivations, professional biographies, life histories and needs, and in this context influences the fields of action of social work.
4. Prostitution takes place in socio-spatial but fluidly interconnected systems (e.g. street, brothel, internet, etc.), which among other things help to determine the methodological approaches of social work.
5. Social work is partly reflected in its understanding of its role and its professional concepts of action in its attitude towards prostitution.
6. It is a field with considerable legal and social contradictions for women, which is also latently transferred to social work, its actors and professional networks.
7. The design of the framework conditions of social work can differ greatly from the point of view of financing and sponsorship.

The number of publications that focus on the related social work is manageable. However, a large part of the literature is not to be found in social work science, but rather in the related disciplines of sociology, jurisprudence or administrative science. An interesting study on the paradoxes of social work's professional action in the prostitution scene is available in Vorheyer and Nagel (2011). Brückner and Oppenheimer (2006) have produced an insightful study of the insights and attitudes of professionals in this field towards prostitution, including interviews with social service workers. Löw and Ruhne (2011) investigated the unseen, unknown and "foreign" experienced field using the example of Frankfurt's train station district with the help of interviews conducted with social workers. Vorheyer (2010, p. 333) tries to identify different role patterns in the field of social work. Howe (2012) reflects the change in the field of prostitution in historical retrospect in the solidary entanglement between the so-called "whore movement" and professional social work. In Brüker's (2011) research, social workers were interviewed as experts on the living situation of older prostitutes. However, it is significant that it is precisely in such qualitative studies that such expert interviews are conducted, in which social workers are asked about the field of prostitution and the situation of the women, but hardly about their own professional situation. When looking at the publications as a whole, it can be stated that basically there are no studies and well-founded analyses from the side of social work science about the professionally active social workers with regard to the framework conditions of their work, the underlying value patterns, the methodological approaches to action and the associated role problems (cf. Albert & Wege, 2011).

2 Organization and Sponsorship of Social Work in a Historical Context

A decisive factor in the analysis of professional social work in the field of prostitution arises in the classification of the historical development, which has a decisive influence on the accompanying establishment and professionalization of the professional field. The following phases can be located in the context of the associated professional social work

1. Phase of the self-help and women's movement in the seventies
2. Foundation phase of the autonomous advice centres in the eighties
3. Establishment and networking phase of the counselling centres and the professionally active social workers in the nineties
4. Differentiation and professionalisation phase of the entire professional field

In the course of the seventies, decisive impulses came from the self-help and women's movement. Social recognition and participation of women in prostitution were part of a progressive and emancipatory social work. This went hand in hand with the founding of universities of applied sciences and the beginning of the academization of social work. The criticism of social conditions and the accompanying processes of alienation within society led to the demand for self-determination and political processes of change. The oppression and discrimination of women was still prevalent despite initial efforts in the educational and professional spheres. The so-called marginalized group theory explicitly included marginalized groups in the focus of social work. The advocacy for the rights of women in prostitution can be located in the context of a comprehensive social change during this time and offered new opportunities for the incipient social movements (Biermann, 1980). This led to the founding and development of a number of autonomous counselling centres in the course of the 1980s. Of central importance was that these projects sought to maintain their financial and conceptual independence from society and grant-giving bodies. Internally, solidarity and partisanship together with women was a central basis for the work of social workers. In a male-dominated society, autonomous spaces were to be created in which women could determine and act for themselves. The steady expansion of the social sector in terms of personnel enabled the establishment and public recognition of the social work associated with it. Innovative social projects became established specialist counselling centres which, although financed by public funds, were able to retain a certain independence. What is significant about this development is that such advice centres tended to be established in large urban centres. The number of counselling centres in this phase was thus manageable and offered opportunities for solidarity-based networking. The joint efforts of the so-called "whore movement" with social work resulted in a series of conferences and projects that both sex workers and social workers organized together. The integration of sex workers into the work of the counselling centres was an essential objective in order to strengthen the common commitment to women's rights. Such efforts were also reflected in the change of the legal basis. The Prostitution Act (Prostitutionsgesetz) of 2002 created the important precondition that the immorality of prostitution was abolished and was linked to the hope of strengthening the professional rights of sex workers. From this point on, an increasing differentiation of the professional field can also be located. Welfare organisations in particular were able to initiate new counselling centres within the framework of new financing concepts, such as EU projects.

The historical development is one of the reasons for the plurality of agencies in this field. The counselling centre "Hydra" in Berlin was founded forty years ago by activists of the prostitution scene and social workers with the aim of increasingly advocating for the rights of sex workers in a social context. In its wake, a number of similarly structured counselling centres were initiated, such as "Madonna" (Bochum), "Kassandra" (Nuremberg) and "Phönix" (Hanover). Despite state or municipal subsidies, these counselling centres are relatively independent and autonomous in their organisational possibilities. However, there are also mixed forms of welfare state or municipal sponsorship, as is the case for the counselling centre "La Strada" in Stuttgart. The social workers working there are employed by the health department of the city of Stuttgart, but work for the counselling centre, which in turn is under the sponsorship of the Caritas Association.

Although the funding of the counselling centres comes primarily from municipal or state funds, the social workers are relatively free of too many value-oriented guidelines in their professional practice and can determine the focus of their work autonomously as far as possible. Although prostitution is regarded as a profession, it is not recognised as such by society. In this context, reference can be made to the offer of career entry counselling, in which the risks of sex work are explained. Social work should primarily advocate for women's rights and provide assistance and care for the social, psychological and physical effects (Reichert & Rossenbach, 2013). However, within this professional field there is also no clear and unambiguous positioning on prostitution.

The independent providers are opposed by a number of sponsorships of the large welfare associations, such as the Diakonisches Werk, Caritas associations and Pro Familia, among others. It must be taken into account here that the values and specific organisational forms of the welfare organisations also flow into the work of the counselling centres. The Diakonisches Werk has its own funding possibilities, independent working groups and a variety of networking and support possibilities for such projects. Although welfare organisations are almost exclusively dependent on state or municipal subsidies and are thus indirectly subject to public control, these can allow a certain autonomy of counselling within the organisation and with minor restrictions. The work is complex when it comes to a public positioning on the topic, offensive press work and the involvement of volunteers. In some cases, the projects are attached to the respective departments, while there are also projects that are subordinated to the management as staff units. It is not uncommon for professional duplication to be possible in such organisational structures. For example, counselling in the area of prostitution is carried out in addition to work in the General Social Service or in the area of conflict with pregnant women. Only a few projects are run by the state or local authorities. Here, it can be assumed that there is a high degree of administrative framework conditions and requirements, which can tend to restrict the autonomous shaping of the profession. Such projects are usually clearly structured and possibly subject to a high degree of control.

In the description of independent, welfare association and state sponsorship, it should only be pointed out in this context that in some cases there are different concepts, financing and objectives. Such organisational structures have a central influence on autonomous professional action, independent design in the area of media and public relations work, the involvement of volunteers and, of course, in the execution of counselling activities with regard to individual case assistance or case management and socio-political work.

3 Social Work in the Social Space of Prostitution

Prostitution ultimately represents an extremely differentiated and heterogeneous area that is subject to a high degree of surveillance and regulation in the public sphere. The "domestication" of prostitution describes the change from public street prostitution to brothel and residential prostitution (Ruhne, 2006, p. 192). In this context, however, it must be pointed out that public spaces as well as the prostitution sector are constantly in motion and changing. In this regard, Dölemeyer (2009, p. 150) refers to an essential differentiation with regard to space: "Prostitution is not the same as prostitution, johns are not the same as johns, and prostitutes are not the same as prostitutes". Just as prostitution changed its spaces or had to change them due to societal pressures, social work also changed the access routes in the spatial context – social work has in a way always followed prostitution on its paths. In this respect, a socio-spatial analysis that includes such differentiations could be of interest for professional social work.

Interesting analyses from a sociological spatial perspective are available for Leipzig and Dortmund, for example (Dölemeyer, 2009). There are a number of cities in which prostitution is found preferentially in certain streets and neighbourhoods and has a long social and historical tradition in this respect, the existence of which goes back in part to the nineteenth century. A number of established structures can be found, for example, on St. Pauli in Hamburg, in the Bahnhofsviertel in Frankfurt, in the "Leonhardviertel" in Stuttgart or the "Dörfle" in Karlsruhe. Here, entire streets are concentrated with the most diverse forms of brothel businesses, running houses and corresponding bars. Such spaces are basically closed systems and are sometimes even symbolically separated from the outside by "privacy screens". The divergence of inside and outside is reflected in the complexity of the field. Social work usually takes on a walking structure here: social work has to establish direct contact with the women. Through the method of street work, the women are visited in their respective spaces and made aware of the counselling possibilities. Counselling sessions cannot always take place in a classical setting of a counselling centre, but also take place in an open space. This can be directly at the workplace, in a bar or in a café. A second area of prostitution is the form of open street prostitution. Depending on the structure, these spaces have grown (e.g. Berlin-Schöneberg), can emerge and expand, change their location for political reasons (e.g. Dortmund-Nordstadt) or have to disappear completely due to public pressure. Social work can have both a coming and going structure here. In some cases, the counselling centres are located in the direct vicinity of the prostitution trade. This immediate proximity enables a fast and flexible form of counselling. In this context, direct access to the women is probably not insignificant. Social work with women in prostitution is also under economic pressure to prove corresponding case numbers. This is much easier to achieve in the immediate vicinity of the milieu than in counselling centres that are further away due to organisational conditions or their location. Berlin, for example, does not have a restricted area and thus does not have large brothels or a concentration on a particular district or street. The social workers are out and about in the entire city and visit the rather smaller brothels. The situation is quite different in Stuttgart, for example, where the counselling centre is located within sight of the milieu.

The change of prostitution from street prostitution to regulated and shielded spaces is also a consequence of increasing upgrading and devaluation processes of city districts and is connected with permanent conflict situations. The space in cities is subject to market economy regulations. Capital and exploitation interests essentially determine political action, because power can arise through these mechanisms. Prostitution, like other forms of social marginalisation (e.g. in the form of increasing impoverishment, homelessness, incipient segregation in the migration sector), becomes an essential feature of social stigmatisation for entire city districts. At this point the role of social work becomes ambivalent. As a form of district and community work, it is used with the aim of enabling new identifications with the district and creating networked social forms. However, social work of this kind is also an essential motor for the redevelopment and upgrading of troubled areas. In this context, marginalized prostitution itself is either partially ignored or even perceived as negative by social work. The residents' need for their district to become more livable, quieter and cleaner becomes part of the action strategy of professional district management. Ultimately, however, it is misunderstood that social work suppresses social problems and makes them invisible. Just as prostitution is not the same as prostitution, social work is not the same as social work: the different fields of work and the methods associated with them also generate different objectives, typifications and understandings of roles, which can not infrequently run completely counter to one another. In some cases, such strategies on the part of an offensive district management are even directed against social work in the field of prostitution. In the course of immigration, offers of the counselling centres are criticised to the effect that they open up the possibility for new immigrants to increasingly settle in the already burdened districts. In this way, such offers are indirectly made jointly responsible for the fact that these difficult problem situations become even more entrenched. This problem-centred view of not being able to live with prostitution in the district often prevents pragmatic solutions from the outset. Social work is faced with the dilemma of having to show understanding for all sides, although this does not correspond to its original intention. To stand up for the presence of prostitution as a perceptible factor in the district, a regulated social work, which also has to keep an eye on the objectives of the financiers, the political guidelines of the municipalities and the partly diffuse legal guidelines of the authorities, only succeeds on the basis of well-founded methodical approaches and a professional self-confidence. Ultimately, this means that social work in the prostitution sector must clearly incorporate political strategies for action in order to counteract increasing gentrification in the big cities with clear concepts. This makes social work in the field of prostitution significant not only for the rights of women, but also in relation to the sovereignty of interpretation and exploitation interests of public spaces in conurbations and cities.

Cross-border prostitution spaces, such as those in the Czech or Polish border regions, represent specific special forms. There, cross-border concepts of social work are necessary and even take place in cooperation with professional colleagues from the respective countries. Another form of prostitution counselling is work in temporary or private residences. Here social work takes place almost exclusively as a form of outreach work. Social workers are forced to seek direct contact with the women. The reactions of the women can be very different and continually present new challenges to the professional role. It is assumed that counselling would usually be necessary but is not always immediately desired. In the form of this contact, it must be determined by both sides whether there is a corresponding need. This requires a high degree of counselling competence, professional self-confidence and strong communication skills.

While direct counselling is possible in the previous forms, the Internet presents a completely new challenge for social work in the field of prostitution. There are indeed limits to the virtual space, especially since prostitution always implies a direct physicality of sexuality. However, the Internet offers completely new forms of prostitution in terms of supply and demand. Depending on the type of prostitution, women can offer themselves at different price levels and book appointments. It is also possible to locate new places of prostitution in an anonymous form. As a result, prostitution is no longer visible to social work and requires new forms of contact, communication and meeting places. The first counselling centres have already adapted to these changes and offer online counselling.

In this context, it must be pointed out that a large number of women do not have a secure place of residence, but are subject to a constantly changing presence and flexibility, both spatially and temporally. Just as the field of prostitution is constantly changing and in motion, so too must be the actors associated with it. The counselling period may only last for a few months because the women move away afterwards. This spatial and temporal uncertainty is an essential part of counselling that professional social work has to deal with. When a system and the target group change, the concepts, methods and ways of accessing the clientele must also inevitably change. This requires a high degree of change and adaptability on the part of social work. A counselling centre that offers its services in a metropolitan area, in a medium-sized town or even in a small town is confronted with completely different framework conditions that have to be constantly taken into account in its work.

4 Professional Role Patterns in an Ambivalent Field of Action

The vague field of prostitution, which is socially taboo and consists more of myths and assumptions, has a central influence on the perception patterns of those involved. As a result, the access routes for social work are naturally restricted. Society and politics have a rather negative attitude towards prostitution. Even the prostitution law created in 2002 basically did not enable recognition as a profession, but rather solidified the marginality of women working in this field. The issues of poverty, immigration and forced prostitution currently make it difficult to have a factual discussion that reaches far into professional social work. In a certain sense, prostitution is respected as a social reality, but it is not aggressively acknowledged. Women in the prostitution business can pursue their activity as long as everything takes place within a regulated framework. Deviations are latently sanctioned or monitored and are thus always subject to a certain discrimination. This societal approach is also a reflection of the framework in which social work operates. The offensive representation of women's rights and interests is limited in the public sphere. For example, it can be a professional challenge for a social worker to lobby the employment agency for a retraining measure or to demand a secure residence status if, at the same time, the authorities demand proof of the previous source of income or career history. Social work thus has to struggle with great difficulties, which has a continuous effect on the methodical approaches to action and their practical implementation.

The approach, confrontation and processing in the context of daily professional practice with these complex framework conditions is in this respect dependent on many factors. This could depend, among other things, on the socialisation during the training, the duration of the professional activity, the professional attitude and the technical competences, the framework conditions of the employing institution (remuneration, time limit, etc.), the regular participation in supervision, the conceptual basis of the activity and the quality assurance. The existence of professional ethical standards essentially shapes the professional understanding of the role of social workers. A typification of the professional role in relation to attitudes towards prostitution has been explored in more detail by Vorheyer (2010). Such typifications cannot be found unambiguously in professional reality, because such positions can change in regards to processes and constantly depending on professional situations. Nevertheless, they can provide orientation about the associated approaches to action, objectives and role assignments. The interface among the different typifications is likely to be much higher in the characteristics of the respective social work and is often overlaid by the positioning and value attitude towards prostitution. Following or expanding on Vorheyer (2010), the following typifications can be assumed, among others:

1. Traditional feminist social workers with a hostile attitude
2. Neo-feminist social workers with an advocacy stance.
3. Liberal-feminist social workers with an ambivalent attitude

The abolitionist position is primarily found in the group of so-called traditional feminist social workers. Although this group recognizes prostitution as a social and societal reality, it is sceptical to partially dismissive of it and is characterized by a rather deficient problem definition of the field of action. This attitude towards prostitution also implies a special form of victimisation. The women concerned are regarded as victims in the milieu of prostitution and are thus subject to ongoing discrimination and exploitation. This social work perspective is ostensibly problem-related: prostitution is a stressful activity with no lasting perspective, insofar as an exit is basically pre-programmed. This also implies the assumption that women tend to work in prostitution involuntarily or have latent exit desires in the process of their activity. However, there is a high level of tolerance and appreciation towards women, prostitution is nevertheless critically questioned or rejected in certain forms. This type is attached to a rather "conservative" feminism, for whom gender relations are characterized by unambiguous and clearly defined power relations. Prostitution is a historically conditioned form of exploitation and reinforces men's need for control over women. This attitude is partly reflected in the "Swedish model" (Dodillet, 2006). In this sense, prostitution as a purchased sexual service represents a special exercise of violence against women. In this understanding, it is precisely these women who require a high level of social, legal and state support measures. Accordingly, prostitution is not a punishable offence, but rather the purchase of prostitution by clients is a violent and degrading act against women. In this respect, it also seems consistent not to prosecute the women, but the men or clients.

The environmental relations of traditional feminist social workers operate in the field of action of open social systems. This includes a form of latent negotiation processes with the social service systems, the specialized authorities and the police. Such social work is part of a moderately acting regimentation and care and thus also takes over partly identical objectives within the prevailing socio-political discourse. Of great importance for such professional systems are financing bases and the associated equipment of the corresponding professional social work. This is also linked to the objectives of purpose-oriented social work, which is to be geared, among other things, to women's wishes to leave the labour market and to their reintegration into the labour market.

In contrast to this is the type of neo-feminist social worker. This emphasizes the advocacy attitude towards prostitution as a profession and advocates the social recognition of the rights of prostitutes. In doing so, the developmental possibilities of women are not viewed in a restrictive way, but in a differentiated way. The women are not only met with tolerance and respect, but also with clear solidarity and support for their actual activity. Prostitution is legally and financially declared as a profession, which takes over an important social function as a form of sexual services. Both sex workers and clients are granted the right to determine the type of sexual service they provide in return for payment in a clearly defined negotiation process. There is a strong demand not to exclude the milieu or to criminalise it without reason. Sex work is an accepted form of socially practiced sexuality. Sex workers should be advised with regard to their activity, among other things, so that they can represent their rights independently. However, this also implies that there are experiences of violence by clients, as a result of which the institutional help systems in particular fail. Administrative policy tends to be experienced as repressive, and conflicts with the police and public order offices are part of this professional understanding of action. In a political sense, social work also sees itself as an advocate vis-à-vis society. In this context, active solidarity with sex workers and the commitment to strengthening their rights in public are consistently advocated.

The type of liberal-feminist social worker is characterised by an ambivalent attitude towards prostitution. In principle, the activity of women is respected, but its effects are met with scepticism. Prostitution does not appear to be a profession like any other, especially since there are no professional protection regulations. The field of action is viewed in a differentiated or processual way, in which there is no clear positioning. There is rather a pragmatic attitude here, which always adapts to the circumstances and life plans of the respective women. Those women who consciously decide for this area and also want to remain in it are therefore not the central target group. Rather, the focus of counselling is on those women who have to struggle with the classic problems such as over-indebtedness, precarious living conditions and psychological and physical consequences. Remaining in prostitution is respected, although clear offers for exiting are also designed. Social work is understood as a long-term accompaniment, in which the social work intervention is completely tailored to the needs of the women. Prostitution can also be understood here as a biographical life model, in which voluntary activity can be assumed at the beginning and in which the women can assess their activity very precisely. In the course of the further employment biography, doubts may arise about the further continuation, in which alternatives are consciously examined. Turning points such as pregnancy, physical illness and psychological exhaustion are occasions for contacting a counselling centre. Depending on the motivation and the availability of concrete and realistically feasible exit offers, social work can intervene accordingly and provide support.

All three typifications represent an attempt to describe which complex and often ambiguous values, role understandings and focal points of action are to be located in the field of professional social work. However, a whole series of central aspects remain unconsidered. There is currently a lack of well-founded research on the extent to which these attitudes are becoming more differentiated. The duration of professional experience, the framework conditions of the employing institutions and the design of autonomous professional action can influence and change corresponding value attitudes in a processual manner. For social workers with different lengths of professional experience, solidarity with the women concerned may be of equal importance, but there are different assessments in the evaluation and value attitude towards prostitution. The generation of social workers or the age cohort to which the respective professionals belong could also have an influence. The demand for political emancipation, critical enlightenment and individual self-realisation play a rather subordinate role in current social worker training as well as in the corresponding social work discourses. Feminist or women-specific discourses are rather supplemented by gender-specific content. In terms of professional ethics, reference can be made in this context to the expansion from the so-called "double mandate" to the "triple mandate". The professional attitude of social work is understood as a special form of human rights profession. In this respect, it can be assumed that a much broader range of professional motivations and attitudes can currently be found in the field of action.

5 Consequences for the Needs of Professional Social Work

On the basis of the present statements on the relationship between social work and prostitution, more questions than answers arise. The existing counselling centres have a long tradition of social work in the field of prostitution and a high level of professional and methodological foundation in the approaches used. There is also agreement to distinguish the different terms such as voluntary prostitution from human trafficking or to clearly separate and define them. Respect for women's dignity and commitment to their rights are at the heart of all social work. However, the plurality of agencies and the different attitudes towards prostitution prevent a certain clarity and standardisation of objectives and strategies for action.

This results in many sub-areas and theoretical assumptions in the relationship between social work and prostitution, which have hardly been researched and described so far. Of particular importance seems to be the reappraisal in the historical context from the perspective of social work. This focuses on the question of how large a part social work played in setting up the independent counselling centres and how structures within the organisation could be linked to external objectives. A systematic reappraisal within the framework of a qualitative study of this "founding generation" of social workers is essentially still pending here. This also includes to what extent and in what form the so-called "whore movement" received support from social work. In the further course of history, it remains exciting to work out to what extent social work was an impulse generator for the establishment of new projects. It can presumably be assumed that social work was the professional authority that first addressed this marginalized area within the framework of social discourse and thus at least clearly formulated the social contradictions. These findings could also be an important basis for the further expansion of counselling centres in order to benefit from previous experiences. Social work must always be thought of and shaped in a historical and socio-political context, in this respect it seems understandable that there are decided differences in the professional socialisation of social work in the 1970s and that in the present. Perhaps a careful analysis in the historical context will even reveal that the struggles and hopes of social workers from the early days hardly differ from those of professional actors in the present.

As has been shown, social work is carried out by different organisations and with different target groups. This will be an important question for the future as to how this affects the focus of action and the methodological approach. The already existing pluralization and differentiation of the field has so far prevented even more commonalities in the professional understanding of social work from emerging. For a well-founded understanding, a comprehensive analysis of all counselling centres and the landscape of providers appears to be indispensable. The attitude towards prostitution seems comprehensible when considering the historical development and the socially contradictory way of dealing with it, but it only allows for an open discourse on the characteristics and components of the social work associated with it to a limited extent. It would be a relevant question for social work science to investigate these differences as well as the similarities from a professional point of view in order to achieve a methodological and theoretical foundation. Here, a qualitative research approach in interviewing social workers working in the field would be informative and could provide important insights into professional identity development. There is also a need for a closer examination of the extent to which the described role patterns of social work apply or whether even further differentiations must be assumed.

Just as it is not known how many women are involved in prostitution, there are also no reliable figures on the number of social workers who offer counselling and assistance at all. Here it must first be clarified how many staff are employed in the clearly defined counselling centres. In addition, there are interfaces between the various social work professions that are more indirectly involved in counselling, such as health offices, migration services, women's counselling centres and street work projects. In fact, only a very small part of the required demand is likely to be covered by the existing counselling structures or professional social work. It is due to the fact that the field itself is taboo and not open to public scrutiny that there is ultimately only little corresponding knowledge available. It is also not known for how many women there actually is a need for social work action and support. This is also due to the fact that the field with its actors and spaces is subject to constant processes of change. The following calculations are not based on concrete statistics, but rather on practical experience with other target groups. If a conservative estimate of only about 200,000 women in prostitution is to be assumed, and assuming that about 20% have a need for counselling either temporarily or permanently, then at least 1000 social workers would be needed in this field, assuming a ratio of about 40 women per full-time position. It can definitely be assumed that there is a latent undersupply of social work in the field of prostitution in Germany. Ultimately, only a well-founded social analysis at the municipal level can provide a well-founded insight into the scene and concretely determine how high the proportion of women is, what the actual needs are and what the resulting need is for professional social work.

Social work is a professional service financed by society, which cannot avoid evaluating its results and checking their effectiveness. Especially in the field of prostitution, it would be significant for the perception of the public and the funding agencies which important and profitable function social work takes on. As is the case for many other fields, quantitative evaluation in the field of prostitution is unhelpful and in some ways counterproductive (cf. Albert, 2006). Classical performance agreements and economized targets are not measurable in social work with women, who are often affected by multidimensional problems. Social work requires autonomous action, in which a lot of time must be available for relationship work. Social work in the field of prostitution requires a high degree of integration of all methods and ranges from individual case assistance to prevention and community work. It must not be reduced to supposed exit programmes, as is demanded almost exclusively by politicians, but is ultimately a counselling service by women for women. It respects different life plans and tries to help individually and situationally, to make professional care possible. In this sense, social work must aggressively advocate both for the rights of women, but also for a sustainable improvement of its own professional framework conditions.

References

Albert, M. (2006). *Soziale Arbeit im Wandel – Professionelle Identität zwischen Ökonomisierung und ethischer Verantwortung*. VSA.

Albert, M., & Wege, J. (2011). Soziale Arbeit und Prostitution – Handlungsbedarf und Entwicklungsmöglichkeiten in einem tabuisierten Berufsfeld. *Soziale Arbeit, 1*, 8–18.

Biermann, P. (1980). *"Wir sind Frauen wie andere auch!" – Prostituierte und ihre Kämpfe*. Rowohlt.

Brückner, M., & Oppenheimer, C. (2006). *Lebenssituation Prostitution – Sicherheit, Gesundheit und soziale Hilfen*. Helmer.

Brüker, D. (2011). *Das "älteste" Gewerbe der Welt. Eine Untersuchung über die Lebenslage älterer Prostituierter. Dortmunder Beiträge zur Sozial- und Gesellschaftspolitik*. Lit.

Dodillet, S. (2006). Prostitutionspolitik in Deutschland und Schweden. Zum ideologischen Hintergrund von Sexarbeit und Sexkaufverbot. In S. Grenz & M. Lücke (Hrsg.), *Verhandlungen im Zwielicht – Momente der Prostitution in Geschichte und Gegenwart* (S. 95–112). Transcript.

Dölemeyer, A. (2009). Gender space und spacing gender – Die räumliche Regulierung von Sexarbeit. In E. Donat, U. Froböse, & R. Pates (Hrsg.), *Nie wieder Sex – Geschlechterforschung am Ende des Geschlechts* (S. 149–183). VS-Verlag für Sozialwissenschaften.

Howe, C. (2012). Struktureller Wandel in der Prostitution. Zwischen Hurenbewegung und Sozialer Arbeit. *Standpunkt Sozial*, (3), 35–47.

Kavemann, B., & Steffan, E. (2013) Zehn Jahre Prostitutionsgesetz und die Kontroverse um die Auswirkungen. *Aus Politik und Zeitgeschichte, 63*(9), 9–15.

Löw, M., & Ruhne, R. (2011). *Prostitution. Herstellungsweisen einer anderen Welt.* Suhrkamp.

Reichert, S., & Rossenbach, A. (2013). "Wir wollen den Frauen Unterstützung geben" Ein Gespräch. *Aus Politik und Zeitgeschichte*, (9), 3–8.

Ruhne, R. (2006). Boulevard und Sperrbezirk – Urbane Ideale, Prostitution und der Kampf um den öffentlichen Raum der Stadt. *Sozialwissenschaften und Berufspraxis, 29*(2), 192–207. (Hrsg. v. Berufsverband Deutscher Soziologinnen und Soziologen e. V.).

Vorheyer, C. (2010). *Prostitution und Menschenhandel als Verwaltungsproblem*. Transcript.

Vorheyer, C., & Nagel, U. (2011). Der habituelle Umgang mit den Paradoxien des professionellen Handelns: Soziale Arbeit in der Prostitutionsszene. In R. Becker-Lenz, S. Busse, G. Ehlert, & S. Müller (Hrsg.), *Professionalität in der Sozialen Arbeit: Materialanalysen und kritische Kommentare* (S. 13–30). VS Verlag für Sozialwissenschaften.

Martin Albert Social Worker (M.A.), is head of the social work program at the SRH Hochschule Heidelberg. His focus is on social work science, professionalization, migration and community work. He has conducted various scientific studies in the fields of street work, district development and prostitution.

Prostitution in the Neighbourhood: Development Concepts for a Design in Public Space

Christiane Howe

1 Introduction

Street prostitution is visible in public space. In most larger cities, this alone makes it a much-discussed topic time and again, as is the case around Kurfürstenstraße in Berlin. Street prostitution has existed at this location since 1885, is a constant part of the neighbourhood, and has been subject to a variety of social, political, and economic changes, such as the urban redevelopment of the West and the transformation of state-owned into private rental property starting in the 1970s/1980s, and most recently the changes caused by the "fall of the Berlin Wall" and the opening and expansion of the EU. Critical discussions about the forms, appearance and effects of prostitution, about different ideas on its handling and design or its prohibition have accompanied it since the beginning in the nineteenth century. What concepts of its design in public space are conceivable today? And how is social work located in this?

Cities can be described as permanent settlements of heterogeneous individuals that are relatively large and densely populated (Löw et al., 2008, p. 11f.). They are "structural, strategic nodes and crystallization points of a society's work organization and consumption" (ibid., p. 13) and are structured with their boundaries via an inside and an outside. Social reality is experienced and interpreted in these urban spaces, own (living/working) places are emotionally occupied. With their density, size and heterogeneity, cities are always at the same time familiar and

C. Howe (✉)
Berlin, Germany
e-mail: howe@ztg.tu-berlin.de

M. Albert, J. Wege (eds.), *Social Work and Prostitution*,
https://doi.org/10.1007/978-3-658-37761-8_3

foreign; diverse and most varied ways of life are possible due to the comparatively low social control. The way of life in cities can thus be described per se as "an achievement" and "an imposition" at the same time (Wehrheim, 2004, p. 21f.).

The common dense use of a public space, a neighbourhood is characterised by a coexistence of working, living and living, which is characterised by a constantly alternating 'with each other' and 'against each other', by ups and downs in living together. Thus conflicts arise again and again at different points in these places of common use. How does this coexistence take shape in a neighbourhood with an existing and visible street prostitution?

In order to record this situation more precisely and to get closer to possible answers, a study was carried out on the street prostitution area around Kurfürstenstraße in the north of Schöneberg in Berlin from April to September 2011 (Howe, 2011).

The questionnaire asked for lines of conflict and approaches to solutions; discussions and guideline-based interviews were conducted with residents[1] and tradespeople, with representatives of various authorities, the police and social work, as well as prostitutes and clients (60 qualitative interviews with 76 representatives).[2] A quantitative questionnaire campaign took place during a public event with approximately 110 people (43 evaluated questionnaires).[3] Site visits as a form of participatory observation were carried out regularly 1–2 times a week during the period, as well as participation in two (professional) meetings and one

[1] In the conviction that social identities, categories and relations are also (re)produced through language, I use the *asterisk as gender distinction and not the generic masculine in order to explicitly identify female and male as well as trans/intersex group members and people who cannot or do not want to fit into the binary gender system. When women, sex workers, etc. are mentioned, those named see themselves as women or are socially regarded as such.

[2] The interviews were conducted according to the method of the problem-centred interview, guideline-supported and content-analytically evaluated. 22 interviews were between 10–30 min long, 38 interviews between 45 min and more than two hours, on average about one hour. Of the 76 total interviews, 32 were digitally recorded with a recorder. 13 conversations were selected, transcribed in full, and 9 were summarized. 4 other conversations were only summarized. 28 conversations were not recorded at the request of the interviewees or due to circumstances and could only be recorded during or after the conversations. In the evaluation, the statements regarding the existing problems and the proposals for solutions from all interviews were taken into account.

[3] Of 110 citizens present, 43 returned completed questionnaires (response rate: 39%). According to conservative estimates, about 20–25% of the residents are only complainers. 75%, however, are interested in information, discussions, solutions and are rather pragmatically oriented. 20–25% interested in further participation.

john action of the Working Group of Healthy Customers. In addition, three workshops (presentation and discussion of the preliminary results with experts, residents, QM representatives – evaluation of the discussions for the final report) were conducted. After the evaluation and analysis, three scenarios were developed and recommendations for action were described.

This study also led to the creation of a 'citizens' exhibition' in collaboration with two artists who have lived in the neighbourhood for many years. The exhibition was set up in buildings as well as in public places and travelled around the neighbourhood in two years during the summer months of 2012/2013 in order to facilitate broader discussions and to collect further proposals for solutions and ideas.

The questions, which are not easy to answer in this context and which persist, can be outlined as follows:

1. Who should and may use the public space in the neighbourhood, which is accessible to all, for what purposes and at what times? Who decides on this?
2. What could acceptable forms of dealing with the shared space and with each other look like for all those involved? What good ideas are there for this?
3. Who speaks for whom? Who is represented?

In addition, in recent years there have been changes and challenges for cities and municipalities triggered by EU enlargements to the East and in particular the immigration of Roma from Bulgaria and Romania. The Roma group is strongly discriminated against there, lack of education, resources and access are the result.[4] Thus, they also look for work in Germany and mostly find it in the low-wage sector. Some of them, women as well as men, also find their income in prostitution. In addition to a visible housing problem (keyword: speculative/junk real estate, overcrowding, usurious rents), these changes became particularly noticeable on the visible street strips. They are a recurring topic of discussion among the residents.

One of the great challenges in this social-urban context – also for the local social work – is not only to support the prostitutes, but also to take into account and

[4] Since January 1, 2014, they have enjoyed full freedom of movement within the EU. There is no evidence for the claim that mainly poverty refugees come from Bulgaria and Romania (Institut für Arbeitsmarkt- und Berufsforschung - IAB Kurzbericht 16/2013). Bulgarian (145,000) and Romanian (262,000) nationals make up about 5.5% of the foreign population in Germany, and Spain and Italy each have one million of them (Bundestag 2013/Drucksache 18/223, p. 4). According to the Federal Government, based on the experience with the previous free movement of workers of the EU-8 states, "it is not to be assumed that significant effects will emerge on the labour market" (ibid., p. 6).

think about the different starting positions, developments and needs of the old and new residents and traders, the changing environment of prostitution as well as the spatial conditions. A guiding idea and vision for this could be an emancipatory one, in the sense of bringing the neighbourhood together in a lively form that enables constructive coexistence.

The conflicts of use, the diverse actors and networks involved show that social work in this neighbourhood, in the social space, must be thought of, oriented towards and acted with – in order to ultimately be able to act in the interests of the prostitutes and with regard to conflict resolution and to remain capable of action.

In the following, after the descriptions of the current situation, the described problems and possible solutions as well as the networks and most important actors in the neighbourhood, the location of social work and the urban context will be outlined.

Brief Preliminary Remarks on the Prostitution Framework

It should be recalled at this point and in advance: Prostitution is the performance of sexual acts for a previously agreed fee. In Germany, it is generally permitted for and with women and men over the age of 18. Every activity and action on the part of the prostitute must therefore be agreed upon and paid for, so that the beginning and end as well as the limits of the encounter are clearly outlined.[5]

As a professionally working service provider he/she generally regulates the process: from the establishment of contact to the negotiation to the concrete design of the intimate communication with the client. On the part of the customer, the encounter can be described as a 'projective staging' of sexual and erotic fantasies, on the part of the prostitute as a professional, thoroughly intimate, but not a personal-private establishment of a relationship. As a rule, the latter is desired neither by the prostitute nor by the client, as it would leave the professional and protective construction comparable to that of a therapy. Thus, prostitutes also consistently give themselves professional names.

The client-prostitute relationship cannot therefore simply be described as a violent relationship. The use of power and violence is not a characteristic of clients, even if there is certainly a willingness and activity to use violence among them, as in partnerships and marriages.[6]

---

[5] In the area of trafficking in human beings (§ 232 StGB), however, there is a protection age limit of 21, which is unusual, i.e. "systematically incomprehensible in the legal framework" (Frommel, 2007, p. 2).

[6] For these crimes – from assault to rape – there would be no need for new laws, as they already exist and only need to be applied (Howe, 2007).

Clients are men (rather rarely women) from all walks of life and educational backgrounds, of all ages. About half of them are married or in a partnership. According to serious estimates, the average share of customers in the sexually active male population is 20%, i.e. about one fifth. That's not the majority of men, but it's 'everyone'. And thus it is men who are as poorly or as well informed about the issue of trafficking as many others.[7]

The described agreement in prostitution constitutes a legally valid requirement in Germany (§ 1 ProstG). Persons from the old EU countries can legally work in all areas of the sex trade, but persons from the new EU accession countries can only work as self-employed persons. Persons from non-EU countries need a corresponding residence status with a work permit.

According to the Federal Government's report on the ProstG, prostitution in Germany[8] is to be understood as an autonomous decision to be respected by the law, but one that is fraught with considerable dangers and risks. "These include, for example, psychological and physical effects on the person concerned. These risks and dangers, however, are not associated with all forms of prostitution to the same extent, but depend essentially on the conditions (emphasis added) under which it is practiced" (BMFSFJ, 2007, p. 7).

Trafficking in human beings for sexual exploitation (Section 232 of the Criminal Code) occurs when persons exploit a predicament or the so-called foreign-specific helplessness of other people in order to bring them into prostitution or to perform sexual acts that exploit them or prevent them from giving up prostitution. People are considered helpless when their ability to act is so limited by their residence in another, foreign country that they are unable to resist working in prostitution.

It is implicitly assumed here that, due to the general conditions, no real consent to work in prostitution can take place or exist, i.e. that every foreign woman is a victim of her circumstances. With this orientation, exploitation refers primarily to the sexual right to self-determination, which is consequently violated here. Fatally, it cannot (any longer) be linked to the right to adequate working conditions, although this would correspond to the reality of many migrants working in prostitution. Most of them have made a conscious decision to do so. Prostitution,

---

[7] The figures are based on six existing studies on johns (Kleiber, 1994; Velten, 1994; Rothe, 1997; Grenz, 2005; Gerheim, 2012). Clients of prostitutes are those who visit prostitutes relatively regularly, i.e. between once a year to twice a week.

[8] Since 1901, prostitution has been officially considered immoral (commercial immorality) and has survived all amendments to the law and political upheavals. It was not until January 2002 that the 'Law Regulating the Legal Relations of Prostitutes (Gesetz zur Regelung der Rechtsverhältnisse der Prostituierten)' came into force, which brought about a change (Schmitter, 2004).

with its low-threshold access, is an opportunity for them to earn a living for themselves and their families (Rabe, 2013). They suffer from the structural, i.e. exploitative and poorly regulated working conditions.

The decision to work as a prostitute is definitely a risky one, but the reasons for this are also to be found in the inadequate protection of legal sexual services under labour and tenancy law (and of course even more so for people working here illegally).

Other elements of human trafficking are deception about the true conditions of the work, abuse of authority and debt bondage. Human trafficking can only be proven if the women concerned decide to make appropriate statements and are available as witnesses.

The causes of human trafficking and its forms of exploitation are manifold. The whole thing is made possible and encouraged by restrictive immigration and labour market policies[9] and the economic disparities between countries. The migration or, more precisely, the labour migration of women (and also men) is due to the precarious situation in the home country.

Women (and men) can become victims of trafficking in various ways. Contrary to the widespread and predominant portrayal in the media, many women are explicitly recruited to work in prostitution in their countries of origin or migrate themselves in an organised manner. However, they then often agree to working conditions that they do not find in the same form in Germany.

2 The Framework: The Neighbourhood

Before the fall of the Wall, street prostitution in the area around Kurfürstenstraße in Berlin was on the outskirts of the city, right on the inner-city border. Today it is a neighbourhood in the centre of the city: south of Potsdamer Platz, not far from Tiergarten, the embassies and the diplomatic quarter, east of City-West with Wittenbergplatz, KaDeWe (Kaufhaus des Westens / "Department Store of the

[9] Regular migration is prevented by restrictive immigration policies and laws. An illegalized or insecure residence status makes migrants vulnerable and exploitable. "This has already been noted and also rebuked by various international bodies. For example, in 2010, the European Court of Human Rights (ECHR), in a case against Cyprus and Russia, found that Article 4 of the European Convention for the Protection of Human Rights and Fundamental Freedoms (ECHR, prohibition of slavery and forced labour) not only obliges states to create sufficiently effective laws to prosecute trafficking in human beings and to ensure victim protection and prevention. Legislation from other areas, such as immigration law, must also be designed in such a way that it does not encourage human trafficking" (Rabe, 2013, p. 19).

West") and Kurfürstendamm, and west of Gleisdreieckpark with the new apartment buildings and new park being built there. Changes are still visible and noticeable in many corners to this day.

This neighbourhood is home to old and new residents, old and young people, families with and without a migration background and with a wide range of educational qualifications and occupations. There are many families with children. A total of almost 26,000 people live in the two areas, around 60% of whom have a migration background. Furthermore, there are four social institutions for families, young people, senior citizens, three schools, two kindergartens, two parishes and a mosque.

Many people who do not live here also work in this neighbourhood during the day. Many run a business. From small cafés, restaurants and bars to a variety of shopping facilities for everyday needs. everything can be found on site. In addition, there are large furniture stores and production sites for the media world.

Structurally, the neighbourhood is also heterogeneous, with Wilhelminian-style old buildings and interspersed newer housing stock from the 1970s and 1980s. Traffic congestion is high due to the high volume of through traffic on three streets – Potsdamer, Bülow and Kurfürsten. And in the middle of it all, street prostitution takes place.

3 Prostitution on the Street

The street prostitution in this neighbourhood is the only one in Berlin that takes place around the clock – in contrast to the situation in the centre of Berlin, Oranienburger Straße. There, it only starts in the evening, the women are all dressed similarly and use rented apartments nearby for the execution of the sexual service. In contrast, the situation in Kurfürstenstraße is far more diverse. The prostitutes alone are very diverse, there are some transwomen (man-to-woman), the so-called 'trannies' and many women with different migration histories working there. Currently most of the women are from Central or South-Eastern Europe, many from Hungary and Bulgaria. Some of the prostitutes work as a sideline, some as a main occupation, others to finance their drug consumption.

The environment also varies. Some of the women work with and others without supportive or exploitative 'pimps'. Around the women exist more or less visible different forms of 'caring' and 'management'. The extent to which pimping is present, even on a larger scale, cannot be precisely determined. The drug-addicted women who prostitute themselves seem to be relatively uninteresting to the pimps. They are attributed with unreliability and high financing needs due to their drug use. Such a pimp-like role is apparently often assumed by drug-using partners or "other male users who are paid to provide some protection during the acquisition process" (Langer, 2010, p. 188).

Some experts say that pimping affects about half of all women around Kurfürstenstraße, some estimate that it is more and that almost no woman works without a pimp. Visible at the moment are mainly Hungarian men, because they move there with their cars (with Hungarian license plates).

There are prostitutes who work rarely, others daily. Some of them are dressed 'professionally', others more in a casual look. They are almost all between 18 and 35 years, only very occasionally underage and rarely older than 50 years. About a third to half of them are only there between 4–6 weeks to three months and then move on. So there is a certain fluctuation.

In the southern part, a small street lined with large trees and old buildings, only the so-called trannies prostitute themselves. They start their work at nightfall and work until the morning hours, daily at least 3 up to (rather rarely) 15 prostitutes. They are mostly from Bulgaria and Latin America and obviously work without pimps. These prostitutes are not infrequently subjected to harassment and attacks by the male youths from the neighbourhood, but often also by the customers.

The street prostitution is delimited and fanned out according to supply and origin. On the one hand, women from the same countries of origin often like to stand together and on the other hand, those who serve a specific demand, e.g. the older women or the 'trannies', each work in their own street sections.

In addition, the prostitutes stay in the cafés of the social projects, e.g. in the women's meeting place Olga and at Neustart e. V., in the six to eight relevant pubs and kiosks along the street strip. There they rest, warm up or use the toilets. These kiosks and bars are also the meeting places of johns, pimps, alcoholics, drug addicts and partly homeless people.

According to estimates of the experts of the social projects and the police, at least 20, on average mostly 40–60 female prostitutes are active every day. Over the whole year, about 200 prostitutes work here.

In Berlin as a whole, according to estimates by the prostitutes' organization Hydra e. V., about 8000 women are engaged in sex work, mainly in apartment brothels, clubs and massage parlours. These 600–800 smaller brothel-like establishments are quite inconspicuously distributed throughout the city and can be found in every district. Only a very small proportion of prostitutes work on the streets.

4 Customers and Places of Execution

Customers come to Kurfürstenstraße on foot, by bike or similar, but mostly by car. Most of them are from Berlin and the surrounding area, but some are also passing through or visiting from Germany and other countries. This is also visible in the different car numbers of the johns. The street prostitution is known nationwide and is mentioned and advertised in relevant brochures and internet forums.

Currently, there are hardly any rooms left where the sexual services arranged on the street can be performed. According to the descriptions of the prostitutes interviewed, a small hourly hotel with five rooms can be rented nearby for 10€ per half hour. At busy times, women have to queue here. It is also relatively run down, it is not always cleaned and there is rarely fresh bed linen. A little further away there is a second, also small hotel, where rooms can be rented by the hour for 15 €. According to the description of the prostitutes, it has much nicer rooms and is run like a hotel, i.e. here the rooms are 'made'. The disadvantage is that it is more expensive and often difficult to find a parking space in this area.

If the customers 'don't want to go to the boarding house' because they don't like it or because they don't want to pay the extra room rent, there is the possibility to go to one of the video cabins of the nearby erotic department store 'Love, Sex Dreams' (LSD). The cabins are described by the prostitutes interviewed as cramped and very warm. Also some of the residents rent rooms to the prostitutes according to statements of experts of the social projects and interviewed residents.

Many of the drivers prefer to perform the sex also in the car. Most of the women working there know the immediate area quite well and have selected quiet and not easily visible places they prefer.

However, due to the neighbourhood's changing location, from being on the outskirts of the city during the Wall era to being part of the central downtown area, these vacant and underutilized spaces are increasingly disappearing. The opportunities to conduct business discreetly are rapidly diminishing. As a result, the execution of sexual services, especially in the summer months, shifts more and more to the quiet, less frequented but inhabited areas and streets.

At night and if desired, the women, according to the description, also carry out the business more or less on the spot in the immediate vicinity, e.g. in the area around the local furniture department store with a large parking lot, in a small nearby park (Magdeburger Platz) or also on the benches behind a church (Zwölf-Apostel), which stands directly on Kurfürstenstraße or between the parking lots under the Bülowbogen, a listed elevated railway between Nollendorfplatz and Bartholdy Park, on which the U-Zwei line runs.

During the day, some of the prostitutes have also visited the two city toilets, which are located in relative proximity, with a customer and used them for the completion of the sexual service. The toilets cost 50 cents, are usually clean and open automatically after 20 min. "Perfect", as one prostitute commented, "the time is quite enough". The business is usually finished after 10 min according to her description, only unfortunately the toilets are closed at night.

Basically, the prostitutes, according to their own description and also that of the experts of the social projects, prefer to go with their clients to an hourly hotel.

For Kurfürstenstraße, an average daily volume of johns can be assumed as follows: Assuming 40–60 prostitutes serving an average of 5 clients per day, there would be around 200–300 johns per day in the neighbourhood. It can be assumed with all caution that there are at least 100, on average 250 and at peak times up to 600 johns per day. The execution of sexual services is provided accordingly.

For an average of 250 sexual services per day, there are ten rooms in two hourly hotels, four video cabins in an erotic department store and two city toilets, which are used for other purposes, in addition to the customers' cars, for which there are no longer any appropriate and discreet parking facilities.

The conflicts of use in this public space are therefore obvious. They result almost inevitably from the location described and are structural, i.e. pre-programmed.

5 Changes and Conflicts in the Neighbourhood

On the one hand, the fall of the Wall changed the urban location of the quarter, it moved from the periphery to the centre, and on the other hand, the places where business could be conducted discreetly disappeared due to the elimination of vacant lots and opportunities. With the increasing building development on the 'edges of the street' and in the neighbourhood, the population is also changing. The execution of sexual services is increasingly shifting to the inhabited areas, especially to the quiet and rather dark side streets directly adjacent to the street prostitution. In addition, in summer, when the weather is good and on some days, the number of women working there increases.

On the other hand, the composition of the prostitutes has changed since the middle of 2000 in the course of the eastward expansion of the EU. More south-eastern European, predominantly Bulgarian and Hungarian women (including many Roma women) arrived and prostituted themselves around Kurfürstenstraße. The motivation and reason for prostituting themselves here is mostly to provide for their children or entire families, some of whom still live in their countries of origin.

Many of the women have already worked as prostitutes there. Here in Berlin they very rarely know their legal situation, have no health insurance, no registration address and usually have little knowledge of the language and the place

This change in personnel was then clearly visible in the neighbourhood. The women's behaviour, the way they advertised and approached passers-by differed. They were more active and thus more obvious. In the meantime there was talk in the media of a "new dimension of shamelessness....". It would have to be laid down by law that certain parts of the prostitutes' bodies are covered" (Berliner Zeitung, 02.08.2010).

Furthermore, loudness and sometimes noisy arguments, especially at night and in summer, were an issue. This new situation also had a corresponding impact on the stands and the situation of the women who had been prostituting themselves there for years. Unrest and confrontations arose.

People living and working around the street prostitution area increasingly complained about negative side effects, especially about the noise in the evening or at night and the pollution on the street as well as on open and green spaces, and about an increased traffic volume due to john search traffic. Furthermore, the accosting of passers-by, as well as some lightly dressed women in the public space and the sexual act performed visibly in the open or in the cars at night led to partial annoyance. Other points were the behaviour of customers, mostly the erroneous approach of local residents and the topic of child and youth protection. "Even kindergarten children are confronted with the negative side effects" (Morgenpost, 17.01.2009). As a result, families with children would feel increasingly uncomfortable in the neighbourhood.

On closer examination, the street prostitution trade with its specific effects on the neighbourhood can be divided into two phases:

The initiation phase is characterized by the standing around of the women, conversations, the approach of the men, the slow driving around and longer stopping of the cars. The consummation phase is characterized by the stopping and driving of the car and the consummation of the sexual service. For sustainable solutions, each of these two phases would have to be considered separately, for example, if one wants to reduce car traffic, these two places should be as close to each other as possible.

In summary, the following problems arise from the above:

- Clash of life, work and living in the district: different people of all ages have conflicting interests at the same time, e.g. working and sleeping;
- Changes due to new buildings and building densification: partly with the development of new resident structures in the neighbourhood;

- Too few suitable places for initiating and carrying out the prostitution business;
- Stable demand for low-cost, low-effort, high-speed sex;
- Change in personnel compositions: Immigration of prostitutes from the countries of South-Eastern Europe since EU enlargement in 2006.

6 Structures and Previous Measures

Activities, efforts and measures on the part of the responsible persons of both districts Tempelhof-Schöneberg and Mitte, the police, the two neighbourhood management teams and representatives of social work can be described as a mixture of reactive-repressive and active-accepting. On the one hand, they are oriented towards the containment and prevention of prostitution in urban terms. This approach is officially and strongly represented to the outside world and is gladly taken up and reinforced by the media. It is an attempt to counter the loudly voiced arguments and criticisms of a few residents and to calm them down with the illusion of a displacement that can be shaped. Possible changes were introduced or implemented in the squares and streets in question. Measures were, among others, with regard to john search traffic: prohibition of passage at night, installation of cul-de-sacs, alternating parking and with regard to the enforcement of sexual services: Pruning of bushes and removal of ping pong tables and fencing of playgrounds. Thus, in public and semi-public spaces, there were phases of selective relief in some areas, but in others this led to increased pollution again. In the warmer seasons, this rises again continuously. On the one hand the prostitution business then becomes more visible again, on the other hand the noise level increases. Here it was laboriously tried to "contain" the whole thing somehow and not to let it simply "get out of hand". With the result that the prostitution in the area moves depending on the location and interventions.

The above-mentioned measures keep the prostitutes on the move to a certain extent, thus creating a meandering form at the edges of the main (core) areas of prostitution. Thus, the women working there disappear in some street sections and reappear in new ones. As a result, the residents there assume that prostitution is expanding and has now reached them as well. This arouses fears of a spreading 'red light scene' and causes further, new citizens to become outraged and complain.

According to statements by prostitutes, they would prefer to go with clients to an hourly hotel or to clearly defined and limited spaces. However, the prostitutes were not asked about the possible starting points for solutions and possible improvements in the neighbourhood. Their perspectives could therefore not be included. The now urgently needed spaces for enforcement in the form of hour hotels were thus neither discussed nor considered, nor can they be planned.

In addition, both representatives of the urban development offices of the district Mitte and Schöneberg-Tempelhof (in a conversation at a citizens' event in the neighbourhood on 29.10.2013) are strictly opposed to any form of permissibility and location of prostitution. They are of the opinion that the prostitution that has existed there since 1885 would finally become entrenched in its current reduced form. Their starting point is the displacement of prostitution; they conceive of any kind of prostitution practice in defined 'spaces', be it in hour hotels or performance boxes, as a brothel-like activity, i.e. establishments. According to their interpretation, these are not permitted in mixed-use areas. This means that the problematic situation in the neighbourhood remains for all concerned. Possible solutions cannot be worked out for the time being and are not in sight for the time being. This would require a different political approach, i.e. a different clear political announcement.

On the other hand, more social, communicative and also mediative measures were taken for the residents, in the form of various assemblies, meetings, citizens' consultation hours and lectures, but also in the form of training courses for the people working in the neighbourhood. Here, information and discussions are used to work and campaign for mutual acceptance and consideration.

An Example: The Language Mediators

Targeted counselling of the prostitutes with mother-tongue street workers led to improvements in phases. The special counselling primarily serves the preventive and basic health care of the women. However, the talks also include information on the neighbourhood and the surrounding area, on appropriate behaviour and clothing. The hoped-for and desired effect was that changes in behaviour would also make prostitution more compatible with the neighbourhood. However, this is subject to significant fluctuations, as women sometimes only stay in phases and constantly change.

7 The Players

Central actors in the neighbourhoods were and are above all the two neighbourhood managements (QM) Tiergarten-Süd and Schöneberg-Nord. Through direct discussions and public debates, they also moderated and guided local negotiation processes (Künkel, 2008), as was the case with the discussions surrounding prostitution. The programme "Socially Integrative City", from which they emerged, aims at the so-called upgrading or stabilisation of disadvantaged neighbourhoods and promotes the activation of residents and the networking of tradespeople, especially with regard to social, political and economic issues and problems. Important work assignments are the activation, networking and improvement of communication among each other as well as the increase of identification with the neighbourhood (Schröder, 2010, p. 52). To this end, a variety of activities have been launched with regard to citizen participation and the initiation and support of local projects. A central goal is also to use this work to reduce the "conflict level" in the neighbourhoods and keep it low.

The attitude of both QMs towards prostitution differs gradually, but can basically be described as ambivalent. On the one hand, prostitution is recognised by both as work and, in connection with poverty and drug use, also defined as a health and social problem. On the other hand, both QMs consider the environment (pimps, customers), the visibility of the women, but above all the effects of prostitution, e.g. through noise and garbage, as a disturbance and burden with regard to the neighbourhood. On the one hand, because it repeatedly causes complaints on the part of the residents, and on the other hand, because in their opinion it impairs the 'reputation of the neighbourhood' and hinders the objectives of their work.

There are also different assessments of the extent to which the prostitutes have a right to be present in the neighbourhood, in the sense of "the prostitutes have always been there", or whether the visibility and the "miserable character" of prostitution is such a "damage to the image" and an "obstacle to development" (Künkel, 2008, p. 181f.) for the neighbourhood, which must be counteracted as far as possible. In the latter assessment, prostitution, i.e. the women with their activity, represents in principle an undesirable or misguided use of space. The taz (German newspaper) also stated that all previous attempts to upgrade the Potsdamer Strasse area as a shopping and strolling mile were endangered by "milieu-related side effects of street prostitution" (taz, 04.12.2007).

A fundamental criticism here is certainly that the 'Socially Integrative City' programme first defined and established the corresponding urban neighbourhoods as disadvantaged and that in the discussion about the implementation of the objectives, i.e. the so-called upgrading or also stabilisation of disadvantaged neighbourhoods, there are obviously also always considerations and discussions about a 'sensible' composition of the residential/working population. This means that not all the people living there are included and a discussion about the social affiliation or non-affiliation or desirability or non-desirability of certain people or groups in the district flares up again and again.

Overall, however, both QMs now assume that the sex trade – in view of the long period of existence on site and a non-existent restricted area ordinance in Berlin – ultimately cannot be displaced and that one must somehow live with it. The consensus and objective that have emerged in the course of the debates in the neighbourhoods are: minimization of disturbances and burdens, especially for the residents, but also for the traders (without taking into account the prostitution business) – in each case oriented to concrete complaints, i.e. to those who bring theirs forward most persistently and without providing a well thought-out overall concept for it.

For the prostitution business that has existed there since 1885, both QMs, which have been active there for over ten years, have not yet developed a concept that understands prostitution as an existing business of the district that could perhaps also have positive effects. Due to the ambivalences and attitudes described above and the objectives of the 'Socially Integrative City' programme, it was seen as an obstacle and a nuisance, rather as something to be contained and suppressed. It was never conceptually considered for the development of the neighbourhood as a business and thus not developed.

In addition, the Berlin police built up prevention teams in the police sections in the mid to late 1990s in order to improve cooperation with the administration, the business community, support institutions, local residents and the gay scene. Today, these teams carry out violence and crime prevention on the basis of a low-threshold conversation offer in the neighbourhood. To this end, there is, among other things, continuous networking with all relevant authorities, institutions and those affected in this area. Through the representatives of the local prevention teams, there is a continuous exchange of experience with residents, the district institutions and the representatives of social work, in particular also through the regularly meeting "Fachgruppe Kurfürsten". This police work on Kurfürstenstraße is basically oriented towards de-escalation. In addition, the Berlin LKA investigates offences such as pimping and human trafficking.

The Kurfürsten expert group has established itself as a very important network of representatives of the social projects and the police, which regularly exchanges information, discusses coordinated and joint intervention possibilities and, if necessary, initiates projects and also offers help. Through this long-standing work and this functioning network, there is good knowledge of the situation on the ground and preventive measures can be developed and implemented accordingly.

In addition, numerous representatives of various senate administrations, the affected districts and, among others, heads of day-care centres and schools as well as committed residents dealt with the issue.

According to the Senate Administration (Berliner Abgeordnetenhaus/Kleine Anfrage, 2008, p. 2), numerous inter-district, inter-authority and inter-departmental discussion rounds have taken place in recent years with all the actors involved and listed here.

In addition, there is an association of Potsdamer Strasse tradespeople, although they are currently rather unconcerned with the issue.

8 Social Work

Social workers are fundamentally and first and foremost committed to the clients, the needs of the individuals, then to the state, the conditions of the state legal system and the current social policy, as well as to the code of social work, which is based, among other things, on human rights.

In the context of street prostitution work, it becomes clear – perhaps more clearly than in other areas of social work – that there are other central fields of work here in addition to the work with the individual clients, the individual, psychosocial care of the individual case. In order to improve the working and living situation of those affected sustainably and beyond mere exit programmes, prostitution with its specific working conditions and the urban context and space in which it takes place would have to be taken into consideration, also in order to be able to further develop and improve the situation with and in the sense of the women who want to work in prostitution. This includes:

(a) to focus on the working conditions of women, which includes not only health but also working techniques, legal principles, etc.
(b) not to shy away from working with the women's environment, e.g. involving their families or even so-called pimps,
(c) working with experts in the field in cross-cutting, interdisciplinary working groups,
(d) work with local residents, possibly local tradespeople, and
(e) to work on one's own positioning in this network as partisan-professional social workers – also with regard to political demands and socio-political designs. The question of self-image possibly arises more strongly here than elsewhere.

On the one hand, the counselling contacts regarding the prostitutes around Kurfürstenstraße show that some of the newly immigrated women are 'uneducated', some are not informed about health and work and often cannot read or write. Furthermore, it is evident that some of the women are in a physically worrying state of health and are under a lot of psychological pressure. Access to the women can only be achieved through the language mediators, and health issues provide a good entry point here. Once a certain basis and trust has been created through this work, the connection to the projects offering care such as Frauentreff Olga, Fixpunkt, Gangway and subway around Kurfürstenstraße can be achieved.

Legal employment status, police registration, health insurance and taxes as well as work techniques, the (family and neighbourhood-related) environment were and are a fundamental part of the counselling work. This also often means that social workers and language mediators accompany them to authorities and offices.

On the other hand, it can be assumed that very few people know about this so-called red light. Prostitution has remained "an area of society about which there is hardly any secure knowledge" (Löw & Ruhne, 2011, p. 115). Even on the street prostitutes rarely come into direct contact with local residents beyond observation. Some of the tradesmen do appreciate them as customers and have contact with the women through them. Most of the people living and working in the neighbourhood perceive themselves as knowledgeable through media coverage. Common images of clients and prostitutes[10] are reproduced here and can hardly be verified. Again and again the connection to criminal organizations, trafficking gangs, the mafia, pimps and big money is made.

As social workers working in this field, it is also important to critically question these images again and again. It is all too easy to follow one's own ideas and attribute them to the clients. In this way, one also constructs certain factual/problem situations with each other, which allow certain problems and (can) exclude others. In this way, certain problems are also (de)addressed by clients, because they need support.

It is therefore not surprising that most people are rather insecure, afraid or react negatively when they come into contact with the people working in prostitution. For this reason, the socio-spatial anchoring, the information and the cooperation with residents and traders around a prostitution area is so important and meaningful. This work can help to break down prejudices and create encounters between people that are still not taken for granted.

In general terms, the demands on the social workers working in the neighbourhood can be described as a kind of community work that goes beyond a job profile regarding the care of women working in street prostitution and an exit orientation. Community work is certainly one of the basic forms of social work. It deals with social intervention in cities and municipalities and is mainly used in

[10] For example, describing clients as old, fat, ugly, moronic and lonely or as insensitive and incapable of relationships as well as sex- and power-hungry. Women, and especially migrants, who work in prostitution are often described as victims of circumstances or violent circumstances, as exploited or as having grown up in deficient/precarious to violent conditions.

'social hotspots' with manifold problems and a corresponding potential for conflict. In principle, the situation around Kurfürstenstraße corresponds to this description and should also be understood in this way. The social workers, however, are responsible for the target group of prostitutes, are mostly understaffed and should nevertheless take up problems in the neighbourhood that are supposedly or actually caused by prostitution and work on a long-term solution with many parties involved. The social workers or professional specialists can be understood as part of this neighbourhood, this community. They are anchored in it and are active across the board in the neighbourhood – also and forced by the circumstances.

Due to the public street prostitution, social work is also a part of the public, and accordingly it is also requested and demanded. At the same time, however, they are not responsible for the community in the neighbourhood and are usually not staffed accordingly.

Due to the design and access of their work, they could act in the role of an active and partisan actor. Furthermore, an important part of their work could be to empower not only the prostitutes, but also the citizens themselves to work for improvements.

Thus, the counselling centres are used as a contact point for complaints by those residents who do not dare to speak directly to the prostitutes or (un)knowingly assume that they do not understand them. This makes the responsible social workers the interface between the worlds and an important actor in the neighbourhood, who can steer information in one direction or the other. This is strengthened by networking with local institutions (authorities, schools, youth centres, churches), but also associations and initiatives. This should be taken into account and more strongly reflected in the work in this area, as it places significantly higher demands on self-reflection, including with regard to one's own attitude towards prostitution.

9 Summary

Basically, it can be stated that both the character and manifestation of prostitution, as well as the attitudes and reactions towards it, are changing in accordance with the structural changes in society. Prostitution is subject to both the common and well-known economic principles and market criteria, as well as the changing ideas about gender roles, i.e. also the gender-specific behaviour of women and men perceived as appropriate, especially in public, and the sexuality(ies) lived. In this respect, prostitution does not stand outside society, but is a part of it.

Prostitution around Kurfürstenstraße can – in view of the existing laws, the long period of existence on site and a non-existent restricted area ordinance in Berlin – neither be prohibited nor suppressed. It is legal and permitted. With regulatory or police measures, action can only be taken against concomitant phenomena of prostitution or against direct violations of the law in this context.

To further limit the spatial possibilities for the prostitutes would have the consequence that the practice of prostitution would be directed to other places and to other quiet, less visible streets in the neighbourhood. Thus, the prostitutes disappear in some sections of the street and reappear in new ones. The residents there then assume, as described, that prostitution is expanding and has now reached them as well. Measures taken so far in the residential environment only provide relief at individual points.

With the reduction of the spaces for street prostitution and the laborious attempt to displace it, there is at the same time also the hope of a "containment of prostitution". Behind this lies the repeatedly mentioned fear that otherwise prostitution would "expand beyond recognition" and could no longer be controlled. This ignores the fact that the prostitution business also follows the laws of the economic market and that the supply of prostitutes is oriented towards the demand of customers. Despite all the exuberant fantasies and prophecies of doom, this demand, the clientele of the clients, is limited. On the other hand, prostitutes who do not generate enough income for their livelihood, e.g. because there are too many women for too few customers, do not stay permanently at such a location and move on. This is also clearly shown by the number of prostitutes, which is subject to fluctuations due to seasons and EU expansions, but which obviously always "levels off" at a certain number. Prostitution could and can obviously not be contained in that sense, but only organized in a "guiding" and negotiating way with all parties involved.

For the prostitution business that has existed there since 1885, however, no separate or joint concept has yet been developed that understands prostitution as a business of the district and that could perhaps also have positive effects. Because of the objectives described, it was never considered in the concept for the development of the neighbourhood as a trade and thus was not developed.

So the question remains: How can the different uses be arranged in such a way that everyone involved can live with them more or less happily?

Obviously there is a lack of spaces for the practice of prostitution, on the one hand for motorists, but also for pedestrians. The question of possible and sensible spaces for the trade must find answers to where and in what framework it should take place? How can there be a sensible demarcation of the different uses? There were a variety of suggestions on this:

1. Urban measures: Positive markings on the street, elaboration and establishment of an eye-catching, comic-like guidance system, work with pictograms, "POP boxes" for cars and pedestrians, establishment of hour hotels (e.g. on the first floor of Wegert House), places for prostitutes to stay (cafés), more city toilets.
2. Creative communicative measures: Education and information for prostitutes, customers and residents/commercial operators, large-scale campaign, round table/working groups, inter-district cooperation, strengthening personal responsibility.

A promising option for action can only be – in the sense of the legislator, who understands prostitution as an autonomous decision to be respected by the law and recognizes it as a gainful activity, and in the sense of the development of the neighborhood, which must include all stakeholders – a recognition and acceptance that includes prostitution as an existing trade. A long-term guiding and controlling can only be achieved by:

- focus on civil and commercial law,
- improved working conditions,
- places in public space as offers and demarcation,
- communicative measures to promote mutual acceptance.

The only design options left are the form of "active guidance" and good offers with regard to the urban conditions, which may or may not be accepted. They would therefore have to be attractively designed if they were to be perceived and accepted voluntarily by the majority of prostitutes and customers.

This approach implies a fundamental change of perspective, taking more account of the civil and commercial aspects, and thus also a change in the content of the debates.

Prostitution is recognised in principle as part of the neighbourhood. An active involvement of all those involved should be made possible for constructive solutions and for an improved design. Overall, all those involved in street prostitution should be seen and perceived as acting independently, as actors who shape and (want to) shape the district in a diverse way. Together with all those involved, i.e. interested residents, prostitutes, traders, customers, "pimps" and experts, it should be considered and worked out in a concrete and constructive manner how the development of the neighbourhood can be shaped together in the long term. The design must be negotiated again and again in a joint process and must also be tied back to the neighbourhood again and again, so that it can develop and consolidate positively. For this purpose, regular meetings of smaller and larger working groups, which work constructively on creative solutions and bring them forward, as well as a regular round table with a larger membership, are central and necessary. Social work should and could play a central and necessary role as "translator" and "hinge" in order to shape the situation in the sense of the affected women to improve their situation beyond individual assistance.

10 (Large) City and Neighbourhood

Ideal-typical ideas of a proper, good and beautiful life in the (big) city emerged in the course of urbanisation processes and the growth of cities[11] as early as the nineteenth century. These ideal concepts are still characterised today by lively, public places where people can relax, recover, meet and exchange ideas in various forms of sociability. These are urban places that allow and encourage encounters between strangers (Anders, 1998), whereby the foreign here by no means refers only to ethnic differentiations. Such places include boulevards, parks, coffee houses, inns and theatres. This resulted in significantly expanded possibilities for action, such as being able to visit "less reputable places" (Wehrheim, 2004, p. 21), such as brothels or sex shops, "without having to fear that the next day everyone would be indulging in gossip" (ibid., p. 21). Behaviour that deviates from the norm, also with regard to gender attributions and orders, can be lived here without fear of sanctions or stigmatisation.

Life in the city thus fundamentally requires a "tolerance towards the foreign" (Siebel, 2000, p. 7) and, on the one hand, expands possibilities and degrees of freedom in the way of life through distance, anonymity and the accompanying lack of control; on the other hand, however, it always generates and is associated with fears and uncertainties, especially with regard to public space.

The urban way of life is permeated by ambivalences, and this tension can be seen clearly in the example of street prostitution. On the one hand, there is an "ideal-typical conception of metropolitan-urban living spaces"; on the other hand, there are "problematized 'deviations' that impair the ideal type and that should therefore be excluded as far as possible" (Ruhne, 2006, p. 193). Prostitution stands for one of these impairments perceived as essential. It has remained a tabooed and often controversially negotiated "grey area" (Laskowski, 1997, p. 80) in urban life

[11] The reasons for this were rapid population growth (between 1816 and 1865 by 60% in the German Empire, caused among other things by the abolition of feudal marriage-restricting regulations and improved hygiene) and, above all, increasing industrialisation (Häußermann & Siebel, 1987, p. 106).

to this day and is subject to special social control, although "there is no actual control problem" (Beste, 2000, p. 247).

It is also prostitution that stands in the way of efforts to improve the civic status of urban neighbourhoods. For example, in Berlin around Kurfürstenstraße, Eastern European prostitutes are shaking the "fragile structure that was painstakingly established through the work of the neighbourhood management and the use of financial resources in the neighbourhood" (Michael Klinnert from the neighbourhood management Magdeburger Platz, Morgenpost 06.10.2007). According to the mayor of Schöneberg at the time, Band, a lot of money was spent "to stabilise this neighbourhood with the help of neighbourhood management and economic development projects. In the meantime, however, one has to deal with a completely new dimension of prostitution" (Morgenpost, 6.10.2007). In his opinion, the most important thing is "to keep the area around Potsdamer Strasse attractive for investors, otherwise the whole neighbourhood is in danger of collapsing" (Bild, 08.11.2007). Not only in Berlin, but also in numerous other German cities, "the urban ideal of 'tolerance towards the foreign' in relation to the field of prostitution is thus merely limited to an acceptance into 'repressed' or excluded tolerance zones" (Ruhne, 2006, p. 193).

Renate Ruhne (2006) also refers to the historical context[12] with which it could be shown, among other things, "that the exclusion of prostitution from everyday public urban life, which is still demanded again and again today, is closely interwoven with the formation and stabilisation of a normative two-gender order, which today shows clear tendencies towards change, but has by no means completely lost its effectiveness" (Ruhne, 2006, p. 204).

Also, the increasing number of prostitutes in the cities was already problematized in the nineteenth century. "Anti-urban resentments" (Krafft, 1996, p. 59), as they were generally widespread, "mixed here with moral concerns and found a lush breeding ground in prostitution. And so it is not surprising that the widespread contemporary discussion of 'fornication' broadened, especially in the 1880s, and finally peaked at the turn of the century. It is no coincidence that the public interest in prostitution shows parallels in time as well as content to the critique of the big city

[12] "The (spatially oriented) control strategies for prostitution that became established as early as the nineteenth century were predominantly aimed at excluding what was stigmatized as 'immoral' from bourgeois moral everyday life. A change in the way prostitution was dealt with was fundamentally linked to structural changes in the rapidly growing cities. However, it also refers to a general 'exclusion' of sexuality or the body from (urban) everyday life that has been increasingly noticeable since the eighteenth century, which shows clear gender-specific differences" (Ruhne, 2006, p. 201 ff.) "and the newly emerging 'bourgeois gender order'.

at the time: At the moment when the attention of contemporaries was focused on the consequences and concomitants of metropolitan development, one of its most characteristic phenomena, prostitution, could not be overlooked" (ibid., p. 59).

Thus, in the twenty-first century, moral concerns are being raised again, especially with regard to children and adolescents. The "limit of reasonableness" would be reached, reports the Berliner Kurier (30.03.2010). Already three years earlier, the Berliner Zeitung wrote that children living in the neighbourhood should not be "burdened by excessively displayed sexuality" (Berliner Zeitung, 29.10.2007) and two years later that this situation is becoming "more and more unbearable" (Berliner Zeitung, 28.11.2009). The head of the women's counselling centre Olga said in the taz (02.06.2009) that it has to be about "a compatible coexistence - not about displacement". This applies both to the residents and especially the families with children, as well as to the prostitutes. "We want everyone to be able to live side by side here" (ibid.).

It is fundamental to note that urban lifestyles always represent both "an achievement" and "an imposition". If the liberating moment is to be preserved and the ideal of the "big city as a place of emancipation" (Wehrheim, 2004, p. 21) is to be realized, this tension must be addressed constructively.

The upgrading and restructuring of public space demanded by QM and many residents can hardly be achieved in the sense of the described ideal with urban planning means alone, since they require a productive "coexistence of different individuals and heterogeneous social and cultural groups and thus the coexistence of different norms" (Wehrheim, 2004, p. 21). This requires constant "adaptations on the part of individuals and permanent processes of negotiation about what is tolerable" (ibid., p. 21). In this sense, it can only be suggested that urban negotiation processes should also be conducted with regard to prostitution and linked to it.

11 In Conclusion

Social contrasts and cultural diversity are close together in cities and can be experienced on a daily basis. This juxtaposition is also regarded by urban researchers as the nucleus of urban culture. The New York urban researcher and social philosopher Richard Sennett (1995) once wrote (about 14th Street in Lower Manhattan) that differences overlap there in one place and that this overlapping of differences is actually the humane centre of the street. The mutual perception and the necessity of arranging oneself make such spaces important learning spaces for living in a modern big city. They require a capacity for conflict and creativity at the same time.

Here, as already described, there are certainly starting points for social work that[13] could be described more as a kind of "structural prevention".

In addition to education, solidarity and support – instead of bans – there is a need for a legal framework (in terms of work, residence and space) in sex work as well as in immigration. These also need to be taken into account, e.g. civil law is also relevant here in addition to the usual social and criminal law.

There is a fundamental and broader need for a perspective that enables individuals to use their potential in a self-determined way. It is necessary to improve living conditions, to involve the environment and thus to strengthen the self-responsibility and self-confidence of those involved, to promote their potential and to empower them. In doing so, it is also important, as Hausschild has repeatedly pointed out, to defend the right to intoxication and passion.

Furthermore, it remains to be stated with Hausschild that the value of a society is measured by how it treats the weakest and what opportunities it grants them to participate and shape it. This is still largely lacking with regard to those working in prostitution.

Social work could see itself in this urban structure beyond individual support and care as a (self-) critical actor, as an important interface and bridge between the worlds and thus take on an important(r) and socio-relevant, not to say socio-political shaping role in these networks. A socio-political, co-shaping role that is oriented towards the participatory, (self-)empowering approaches also in the sense of common and structural approaches to solutions.

With Renate Mayntz (1997), however, one must conclude and restrictively state here that the respective steering action, e.g. of politics, administration, police and also social work, is in each case only a part of the social process, "which interferes with many other sub-processes and thus contributes to social change", without, however, being able to steer it in the sense. "At the level of the system as a whole, no steering takes place, but only structure formation and structural change. This means that there is steering in the functionally differentiated society, but no political steering of society" (Mayntz, 1997, p. 286). The planning and steering actions of the various actors thus represent merely one impulse among many others that influence the developments of cities – even if some people would certainly wish it otherwise. All the more desirable here would be a strengthening of 'professional interference' on the part of social work.

[13] According to Dr. Hans Peter Hauschild, 1990 to 1993 member of the board of directors of the Deutsche AIDS-Hilfe, based on the concept of the World Health Organization (WHO), according to which health requires framework conditions that enable individuals to use their healthy potential in a self-determined way.

References

Abgeordnetenhaus Berlin/ Drucksache 16/12 696. (2008). *Umgang mit Straßenprostitution und Begleiterscheinungen im Gebiet Kurfürstenstraße*. Kleine Anfrage des Abgeordneten Björn Jotzo (FDP) und Antwort. 16. Wahlperiode. Berlin.

Anders, G. (1998). *Stadt der Öffentlichkeit*. Haag+Herchen.

Beste, H. (2000). *Morphologie der Macht. Urbane „Sicherheit" und die Profitorientierung sozialer Kontrolle*. Leske und Budrich.

Bundesministerium für Familie, Senioren, Frauen und Jugend (BMFSFJ). (2007) *Bericht der Bundesregierung zu den Auswirkungen des Gesetzes zur Regelung der Rechtsverhältnisse der Prostituierten*. (Prostitutionsgesetz – ProstG).

Frommel, M. (2007). *Schutz der persönlichen und wirtschaftlichen Bewegungsfreiheit von Prostituierten*. Vortrag zum Strafverteidigertag am 24. März 2007 in Rostock.

Gerheim, U. (2012). *Die Produktion des Freiers. Macht im Feld der Prostitution. Eine soziologische Studie*. Transcript.

Grenz, S. (2005). *(Un)heimliche Lust. Über den Konsum sexueller Dienstleistungen*. VS Verlag für Sozialwissenschaften.

Häußermann, H., & Siebel, W. (1987). *Neue Urbanität*. Edition Suhrkamp.

Howe, C. (2007). *Männer(bilder) im Rahmen von Prostitution – die Konstruktion des Freiers*. In J. Luedtke & N. Baur (Hrsg.), *Was macht den Mann zum Mann. Beiträge zur Konstruktion von Männlichkeiten in Deutschland* (S. 239–263). Budrich.

Howe, C. (2011). *Nachbarschaften und Straßen-Prostitution – Konfliktlinien und Lösungsansätze im Raum rund um die Kurfürstenstraße in Berlin*. Im Auftrag des Bezirksbürgermeisters von Tempelhof-Schöneberg, unter Mitarbeit von Milena Sunnus..

Kleiber, D. (1994). *Prostitutionskunden. Eine Untersuchung über soziale und psychologische Charakteristika von Besuchern weiblicher Prostituierter in Zeiten von AIDS*. Nomos.

Krafft, S. (1996). *Zucht und Unzucht – Prostitution und Sittenpolizei im München der Jahrhundertwende*. Hugendubel.

Künkel, J. (2008). Das Quartier als revanchistische Stadtpolitik. Verdrängung des Sexgewerbes im Namen eines neoliberalen Konstrukts. In O. Schnur (Hrsg.), *Quartiersforschung zwischen Theorie und Praxis* (S. 169–192). VS Verlag für Sozialwissenschaften.

Langer, A. (2010). Auffallen ohne entdeckt zu werden. Interaktionen von Prostituierten und Freiern auf dem ‚Drogenstrich'. In T. Benkel (Hrsg.), Das Frankfurter Bahnhofsviertel–Devianz im öffentlichen Raum (S. 183–208). VS Verlag für Sozialwissenschaften

Laskowski, R. S. (1997). *Die Ausübung der Prostitution. Ein verfassungsrechtlich geschützter Beruf im Sinne von Art. 12 Abs. 1 GG*. Peter Lang.

Löw, M., & Ruhne, R. (2011). *Prostitution. Herstellungsweisen einer anderen Welt*. Suhrkamp.

Löw, M., Steets, S., & Stoetze, S. (2008). *Einführung in die Stadt- und Raumsoziologie* (2. Aufl.). Barbara Budrich.

Mayntz, R. (1997). *Soziale Dynamik und politische Steuerung: theoretische und methodologische Überlegungen*. Campus.

Rabe, H. (2013). *Menschenhandel zur sexuellen Ausbeutung in Deutschland*. In Das Parlament/ Beilage Aus Politik und Zeitgeschichte Nr. 9, Thema: Prostitution Zugriff am 25.2.2013. http://www.das-parlament.de/2013/09/Beilage/003.html. Berlin.

Rothe, A. (1997). *Männer, Prostitution, Tourismus. Wenn Herren reisen...* Dampfboot.

Ruhne, R. (2006). Boulevard und Sperrbezirk. Urbane Ideale, Prostitution und der Kampf um den öffentlichen Raum der Stadt. In Berufsverband Deutscher Soziologinnen und Soziologen e. V. (Hrsg.), *Sozialwissenschaften und Berufspraxis* (SuB). 29. Jg, Heft 2 (S. 192–207). VS Verlag für Sozialwissenschaften.

Schmitter, R. (2004). *Prostitution – Das älteste Gewerbe der Welt? Fragen der Gegenwart an die Geschichte* (Hrsg. Bremer Frauenmuseum). Schardt.

Schröder, C. (2010). *Akteure der Stadtteilentwicklung – Wie Verwaltung, Politik und Bürgerschaft Beteiligung definieren*. Oekom.

Sennett, R. (1995/1983). *Verfall und Ende des öffentlichen Lebens. Die Tyrannei der Intimität*. S. Fischer.

Siebel, W. (2000). *Urbanität als Lebensweise ist ortlos geworden. Die gute Stadt als Utopie von Freiheit, Toleranz und Unabhängigkeit hat ausgedient*. In Frankfurter Rundschau vom 29.07.2000, S. 7.

Velten, D. (1994). *Aspekte der sexuellen Sozialisation. Eine Analyse qualitativer Daten zu biografischen Entwicklungsmustern von Prostitutionskunden*. Unveröffentlichte Dissertation.

Wehrheim, J. (2004). Städte im Blickpunkt Innerer Sicherheit. In *Aus Politik und Zeitgeschichte*. 25.10.2004, (S. 21–27) Bonn.

Christiane Howe holds a degree in sociology and is a research assistant at the Institut für Europäische Ethnologie at Humboldt Universität in Berlin. Her work focuses, among other things, on (labour) migration, urban space research and gender relations. She has led numerous research projects on the topic of prostitution. She is also a freelance author and speaker.

Drug-Using Sex Workers Are Service Providers: A Change of Perspective in Social Work in the Fight Against Sexualised Violence and Exploitation in "Drug Prostitution"

Kathrin Schrader

1 Introduction

Sexualised violence against women is a social reality, but similar to the discussion about spaces of fear, women are thematised as victims, the socially anchored violence disappears behind the victim discourse and is denamed.

> "Sexualized violence is seen as an individual disorder either on the part of the perpetrator or the victim, or as a terrible blow of fate. Dealing with experiences of violence thus becomes the personal responsibility of women again" (Brenssell, 2012, p. 210)

The gruesome images of young girls being sold (Meletzky, 2012; Meyer et al., 2013) and forced into sex understandably cause horror. As a result, the social demand for a ban on sex work is becoming louder again. The habit of immediately linking criminal practices in this specific occupational field with a ban on it shows mental laziness and inconsistency, since no one wants to ban the profession of seamstress after the criminal violations of the most fundamental safety standards in the textile factories of Bangladesh.

K. Schrader (✉)
Frankfurt am Main, Germany
e-mail: schrader.kathrin@fb4.fra-uas.de

M. Albert, J. Wege (eds.), *Social Work and Prostitution*,
https://doi.org/10.1007/978-3-658-37761-8_4

It is fatal and dishonest to instrumentalise the public through the targeted use of cruel images and then to demand the tightening of laws on the basis of the consternation generated and then to legitimise this with reference to social opinion. The victims themselves are not helped by such activities, but they are re-victimized and become the target of state security arrangements. A society that wants to intervene on behalf of the victims of exploitation and abuse must take these people seriously and empower them as subjects in every situation. It is the lawless status of illegalized sex workers and not the sex work that promotes crime.

When we speak of victims, we often mean subaltern status groups, i.e. people who do not appear in the hegemonic discourse with their wishes and demands. In the sense of the supposed good, debates and decisions are made over their heads. Irrespective of the fact that this is often the most effective option, indeed sometimes the only practicable one, the approach must be seen as colonization (Müller, 1986), in order to thereby sharpen the focus on eliminating the causes of violence and exploitation for those "affected", rather than merely treating their symptoms.

Explicitly when it comes to the discussion whether sex work is a profession like any other, whether the activity can be defined as work at all and has to be tolerated by society, (feminist) minds and discourse apologists are divided. Some know exactly what is good for women and others believe that the "people concerned" can speak for themselves. While the former speak about women, the danger in the opposing position is to neglect relations of violence and domination and to underpin the "anything goes" of neoliberalism. That, at any rate, is the accusation that proponents of sex work have to contend with. But even within sex work advocates, opinions differ when it comes to the sexual services provided by drug-using sex workers. Recognition of the concept of work is contested, although many of my interviewees refer to it as work:

> "Yes of course it's work. Definitely. I had the discussion with someone the other day, he is a baker. (...) He said he works hard. And I'm making such a fuss and so on. I say, do you think I don't work? I say, I run around here for hours sometimes 24 hours, I'm glad when I can change shoes, I have thick feet, I've talked my mouth off, I've got some nerve and psychologically, how I somehow get hold of the one, so that I get hold of him. Sometimes it is quite nice and o.k. and then you like to talk but sometimes you really have to listen to some weird shit, I'm sorry, but then it's really like that I have to ask, what does he actually want to tell me now and you have to stay nice and friendly and yes, there is a lot of psycho-, yes it's like being a therapist here sometimes, so..." (Schrader, 2013, p. 232)

This article aims to show that it is important to recognize the sexual services of drug-using sex workers as work, in order to be able to scandalize exploitation and sexualized violence in a work environment and to empower the women themselves.

First, I will describe the violence against drug-using sex workers. Then I will explain why the concept of work in sexual services is necessary for drug-using sex workers in order to denounce the tendencies towards dissolution of boundaries and alienation in this area as well. In the final chapter, I will show which perspectives social work should take with regard to the intertwining of drug use and sex work in order to support self-empowering strategies against sexualised violence.

2 Violence Against Drug-Using Sex Workers

A common justification that "drug prostitution" is not compatible with human dignity is derived from the fact that "drug prostitutes" are significantly more often and massively affected by violence by clients. Kerschl comes to the following conclusions in the study on risks on the drug street:

> "Violent assaults by clients are part of everyday life on the drug street. Overall, drug-addicted women and girls are exposed to a higher level of violence (...). Drug withdrawal and the conditions of the street prostitution harbour an increased risk of becoming victims of abuse and sexual violence. Some clients deliberately exploit the emergency situation of women and girls who are under pressure to procure drugs: They try to push down the prices, demand sex without a condom, humiliate the women or expect unusual sexual practices" (Kerschl, 2005, p. 117)

Zurhold and Kuhn refer to the same results in a study on crack use by young women in the Hamburg drug prostitution scene:

> "The drug street prostitution in particular is a place where everyday violence against girls and women is widespread. (...) Above all threats, but also physical attacks and rapes have already been experienced to a strong extent at least by the girls and women" (Zurhold & Kuhn, 2004, p. 250)

Mitrovic also states for Hamburg that the women in street prostitution have almost no protection compared to the women in other forms of prostitution.

> "Unlike in the well-organized milieu on St. Pauli, drug-addicted prostitutes in St. Georg are often exposed to violence by clients" (Mitrovic, 2004, p. 45)

Another important study in the German-speaking area on the subject of violence and "drug prostitution" is the work of Egartner and Holzbauer (1994). The authors develop their definition of violence on the basis of interviews with female drug users. For Egartner and Holzbauer it is certain that "addicted" women are often exposed to massive experiences of violence in their childhood and also in adulthood (Egartner & Holzbauer, 1994, p. 192).

On the basis of their research findings, they have decided not to limit themselves to the concept of violence, but also to use that of boundary transgression when describing the causes of injuries and traumas, since only victim-related experiences are linked to the concept of violence (ibid.). They elaborate different levels of boundary transgression, which they separate analytically, but they also state that they are intertwined:

- Physical violence,
- Sexual violence,
- Boundary transgressions towards oneself (ibid., p. 193).

The researchers understand physical violence as the physical attack of one person on another, whereby the attacked person is usually clearly inferior. The violence is therefore sometimes experienced as life-threatening (ibid., p. 194). To explain sexual violence,[1] Egartner and Holzbauer first cite sexual harassment, a sexually directed attention or attention that is exercised against the will of the desired person (ibid., p. 205).

However, it is difficult to define such a violation of boundaries in the prostitution milieu as sexual harassment. Visible desire and sexualized looks, words and touching are part of the setting of prostitution on the part of the customers as well as on the part of the service providers.

Rape as an extreme form of sexual violence is a massive physical assault (ibid., p. 207), which often takes place in a close relationship, such as the family or the circle of friends and colleagues. Sexual violence that happens in a closer social environment is still taboo and the victim is often attributed to complicity. As this exacerbates an already shameful situation, perpetrators often go unreported. Butler describes the assumption that a woman herself is to blame for rape as a premeditated labelling of an act:

[1] I must neglect the discussion of "sexual abuse" in this text for reasons of space. The topic requires explicit discussion. Basically, I agree with Egartner and Holzbauer: They distance themselves from the term "abuse" because there can be no "misuse or use" of girls and women of whom they are not victims (Egartner & Holzbauer, 1994, p. 209, footnote 63).

> "Even though gender appears here as a category, this is not a mere representation, but rather a principle of production, a principle of intelligibility and regulation that enforces a violence and rationalizes it after the fact. The same terms used to explain rape simultaneously enact rape, thus acknowledging that rape was already there subliminally before it took on the empirical form of a criminal act" (Butler, 1993, p. 55 f.)

The reporting behaviour among sex workers is even lower than among women in the majority population, because the rape of a sex worker is usually treated as a "work accident". If the woman also uses drugs, it is assumed that she was not in her "right mind" and had recklessly provoked the assault. Furthermore, because of their criminalization, drug-using sex workers perceive the police as part of the repressive state apparatus that persecutes them and from which they must be wary. They therefore do not expect it to protect them (Egartner & Holzbauer, 1994, p. 208). Similarly, the stigmatisation of prostitution also affects reporting behaviour, as this is often accompanied by prejudice and devaluation on the part of the police, and sex workers in this environment feel ashamed to report the violence and humiliation they have experienced (Kavemann et al., 2007, p. 17).

Sex workers complain about this state of affairs and describe that discrimination and social contempt are linked to the view that sex workers cause their own misfortune. "'What's she doing getting into a car alone with a strange man!' Or: 'If she goes to such dark districts, she shouldn't be surprised ...'" (Gerstendörfer, 2007, p. 129).

> "(...) even if I get in the car and something happens and you go to the police and they say yes, why did you get in? It doesn't matter whether you got in or not, they can't do anything you don't want them to do. But they won't be punished. Then they say, well, it's your own fault for getting in. That's what they told me at the station. You didn't have to get in. That's bullshit" (Schrader, 2013, p. 218)

Gerstendörfer writes that at this point cause and effect are confused because society and politics have actively created the framework conditions through restricted areas in which violence against women can be committed (Gerstendörfer, 2007, p. 129). She refers to the myth that it is not possible to rape sex workers because they are always ready for sex (ibid., p. 132 f.). However, sex workers experience violence in the workplace just as negatively as other women (ibid., p. 133). Experiences of violence are always to be evaluated independently of the profession practiced.

Therefore, the term "forced prostitution" is also inadmissible. This term inadmissibly mixes sexual violence and sexual service. If a prostitute is forced to perform sexual acts through violence, she is a victim of crime and is not engaged in work (ibid., p. 138).

This approach by Gerstendörfer is an important basis for an emancipatory concept of work, also for drug-using sex workers. Without this, no awareness of the injustice done to them through violence can develop.

Increased state intervention against "drug prostitution" and repression against "drug prostitutes" will not lead to more protection against border violations and violence, since only restrictive legislation disenfranchises women through criminalization and enables clients to violate human dignity with impunity. However, I am not concerned with a general criminalisation of the clients, but with making the constellation of problems transparent. There is obviously a demand market where people want to live out their sexual power fantasies and/or compensate for their deficits in terms of self-worth. You can find that sick and repulsive, but it is the reality. Its "victims" are always the weakest members of a society, i.e. people who find themselves in a (material) predicament and are endowed with few or no rights. Policymakers could effectively address these issues in order to protect the human dignity of women. Regulating "drug prostitutes" or criminalizing the clients only pushes the problem further into illegality, where the vulnerability of women can be exploited even more.

Everyday violations of boundaries through violence characterise the life situation of drug-using sex workers:

> "Yes of course, on the scene I also have violence and rape, and I don't even know how I even got through all that psychologically so that I still remained the way I am" (Schrader, 2013, p. 249)

Sex work is an activity in which the risk of becoming a victim of violence is very high. Nevertheless, the equation of violence and sex work must be contradicted, as such a view is not productive. It is important to analyse the connections between the reality of life and the experiences of violence in a differentiated way.

> "Violence in the lives of prostitutes is not identical with violence in prostitution. Basically, a distinction must be made between violent assaults associated with the practice of prostitution and experienced violence outside of prostitution" (Leopold & Grieger, 2004, p. 19)

The third level of violence is "transgressing boundaries towards oneself". The women interviewed by Egartner and Holzbauer experienced these frequently and in a variety of forms (Egartner & Holzbauer, 1994, p. 214). People whose boundaries have been frequently violated lose a sense of their own and others' boundaries, according to the authors.

For Egartner and Holzbauer, "addiction to 'hard' drugs, which is extremely self-destructive in its physical, psychological and social effects" (Egartner & Holzbauer, 1994, p. 214 ff.), is the most extreme form of borderline violation towards oneself. This statement, which is only correct in principle, must be clarified to the effect that drugs are consumed not because of their self-destructive effects, but as an aid in coping with problems and thus ensuring the reproduction of the labour force. The negative effects are thereby knowingly accepted, but they are only the side effects of a prohibitive drug policy, which leads to drugs of poor quality being sold at high prices. Together with criminalisation and the repression interwoven with it, consumption has extremely negative physical, psychological and social effects, which are then falsely negotiated as pathological self-destruction. The drug use of my interview partners is also not to be interpreted as targeted self-destruction, but rather as an aid in the struggle for survival, which due to the restrictive and punitive structures can cause social or physical death (Schrader, 2013, p. 277).

> "Concomitant and consequential health damage of illicit drug use is thus less substance-related (if the substances are in a pure state and not, as is common on the illicit drug market, laced with harmful substances), but primarily a consequence of criminalized consumption conditions and contexts of use" (Gerlach, 1993, p. 3)

"Drug prostitution" takes place in an environment marked by violence, discrimination and persecution. Nevertheless, the women I interviewed manage to resist and in doing so they resort to their dignity and honour, the last resources they have left in the situation of distress, even if they have already been violated countless times:

> "And I've always said that dignity is inviolable and that's the way it is. And that's the, I don't know, I don't let myself, even if I'm a drug addict, I want to be able to look at myself in the mirror despite everything. Even if some things didn't go well as they should have, but that's the way it is and that's good. That's part of it somehow but I don't let my damn pride be taken away whether it's a cop [policeman K.S.] or a client. And I think that somehow keeps me together" (Schrader, 2013, p. 231)

3 "Drug Prostitution": Extremely de-Limited and Alienated Work

The alienation of labor is a social phenomenon that has found extreme expression in the activities of drug-using sex workers. Ward writes about this:

> "The lives of sex workers are often hard and dangerous, and not least because they are criminalised and subject to repression, leaving sex workers at the mercy of abuse by pimps and clients. Many sex workers are unhappy with their work and would leave it if there were real alternatives. Yet it is a form of alienated labour like the others under capitalism" (Ward, 2007, p. 7)

Kappeler relates drug use to this and writes that in the developed industrial societies of the present, the instrumentalization of people has assumed an extent in which for many people the contradiction between required rational self-functionalization and permanently violated sensual needs is no longer bearable (Kappeler, 1998, p. 58). People tried to regain the permanently withheld sensual qualities of experience and to compensate for the destructive experiences (ibid., p. 59). This happens in different forms, and the use of illegalized drugs is one of them.

"Drug prostitution" is a work in which alienation as a social phenomenon becomes effective in its most extreme forms. The analysis of the interviews I conducted also proves this with regard to the structural delimitation of work.

> "The worst thing I find is that you wear yourself out physically and sometimes you don't get to rest, and when you do get to rest, there's no opportunity or no place where you can lie down. In the hotel it is also like that, if you have not paid the full price, then you are not allowed to stay there" (Schrader, 2013, p. 243)

> "I have now had to pay 200 euros just because I asked someone for a cigarette. I asked for a cigarette and the police said I asked him to go to the room. And it wasn't like that. I really only asked him for a cigarette and that's all there was to it. So they wrote me up for that. And then I had to pay 200 euros. Only, if you don't have the money, then you have to work. What else am I supposed to do" (Schrader, 2013, p. 216)

The alienation becomes visible in the extremely long working hours that have to be worked due to decreasing numbers of customers in order to be able to earn enough money to pay hotels and fines as well as the high drug prices. Drug-using sex workers work alienated because their bodies are extremely exploited to the point of the boundary violations described above. It takes a lot of strength for them to defend themselves against the violating invocations and exploitative structures in order to be able to maintain their professional ethos and self-respect.

> "It has become a matter of course here that you work without a condom. Or for the clients, I will say, for the clients it is a matter of course, because I often experience that when I go upstairs with them and they say, how? With a condom? that I think, huh, that can't be. [...] So that's the way it is, I could really get a certain clientele, they don't care at all whether a woman is sick or not and yes no idea. I mean, I work with condoms, because I just got this hepatitis C, that's enough for me. [...] Well, I do have principles for myself, that is just not without a condom, no anal intercourse and just what comes too close to me, but I can only decide that at the moment. That is, yes, it is difficult to describe. So there are regulars who I let get closer to me than maybe some guest who only crosses my path once. So I can say that I don't have such principles, it always depends. But I do stick to the principle: not without a condom" (Schrader, 2013, p. 212 f.)

For some of the women I interviewed, sex work is always closely linked to consumption. Drugs serve as a tool to cope with work in which the entire personality (body and psyche) becomes a commodity and in which there is no structural network to fend off the extreme impositions. The women are thrown back on themselves and have to resist the dissolution of boundaries on their own. On the street, most women work independently, i.e. without a pimp. However, this apparent freedom only exists because the earning opportunities are so poor. It goes hand in hand with considerable risks in terms of client violence and leads to poor working conditions. This also includes the fact that there is no structural safety net in the sense of occupational health and safety.

> "[...] am all alone. I have also never had one here either. I always thought they were friends, but they always took advantage of me. When I had money I was good, as soon as I had no money they didn't know me again, I could do without that. I'd rather stay alone. [...] No cohesion, not at all [...] Not at all. Everybody thinks only of himself and tries to get along somehow. There used to be, but that's long gone. [...] Yes, because most of them just don't care anymore. I think they just want to get a customer, they don't notice everything else around them. They just want to make money fast, no matter how and if someone gets in or not. If someone finds out about something, they just say to each other, you're not getting in here or there" (Schrader, 2013, p. 223)

The situation plays into the hands of the clients. Although the women claim to work professionally, the power position of the clientele keeps pushing the boundaries towards complete availability.

For drug-using sex workers, it is important to take care of their own bodies, including an attractive appearance. All interviewees stated clearly defined working standards, which should be observed by the service providers as well as by the clients.

> "Well, I don't understand men anymore sometimes. It wasn't like that before. And I can't think of anything at all. Sometimes I think, should I stop or should I keep going. If I, because I never do without a condom, I don't do it. Someone can give me 5000,- and I won't do it. I don't know if they've already done it with someone without a condom or what. No, I'd rather think about my life. My life is a thousand times more important to me than getting AIDS or some other disease somewhere. So I can think of absolutely nothing more" (Schrader, 2013, p. 264)

This formulates terms and conditions based on the professional ethics of service providers and on ethical dealings between business partners.

Among drug-using sex workers there is a fund of knowledge about professional work and they are aware that they are providing a service and not selling their bodies.

> "Because I don't want to have my dignity and my pride taken away from me. I don't know, I'm not, I don't know if it's my stubbornness or what, but hey, if some guy comes and thinks he can pay with drugs, no thanks, I can get that myself. I have to pay for my room, I'm hungry, he also wants me to look good and not stink and what like where, I have to buy that somewhere to pay. And I don't want him to get a head [crack K.S.] or something like that no, no, no, I don't know, I let him, just because he, wag, wag, he buys a service but not me" (Schrader, 2013, p. 226)

The professional self-image of drug-using sex workers is no different from that of other sex workers. However, this professional attitude is repeatedly questioned or prevented. Drug-using sex workers are still treated as "declassed". This classism can be explained, among other things, by Michel Foucault's dispositif of sexuality. According to Foucault, since the mid-eighteenth century, the technology of sex was a means for the bourgeoisie to give itself a body "to be cared for, protected, cultivated, protected from all dangers and touches, and isolated from others so that it would retain its own value" (Foucault, 1995, p. 148). The bourgeoisie gave itself a sexuality, and from there it created a specific "class body" with its own health, its own hygiene, its own offspring and its own "race" (Foucault, 1995, p. 149). The classes exploited by the bourgeoisie were not granted sexuality for a long time. Sexuality is thus bourgeois in its origin, and through its successive displacement and transmission it led to various class effects (ibid.).

This explains the refusal to recognize sexual service as work. In the line of argumentation of the bourgeois sexuality dispositif, it is still only carried out by "socially marginal" women with their polygamous, bustling sexuality, who also do not need any qualification, since they master sex qua their origin.

Through this excursus, it should become clear that sex work has a classist structure along relations of domination and that domination and exploitation are thus perpetuated. Social and economic conditions force women to work in a field of sex work that is characterized by extreme conditions of exploitation. Therefore, it is all the more important to qualify women in their work, because sex work does not take place outside the capitalist system.

> "Commodities have both use value and exchange value. The use value of prostitution is the satisfaction of the client's desires, the provision of sexual pleasure. The exchange value is the social labor contained in the commodity, that is, the physical and mental labor contained in the provision of the sexual service. It is equivalent to what a sex worker needs to reproduce herself under the socially average conditions for this industry" (Ward, 2007, p. 3)

For drug-using sex workers, it is also true that they can hardly cover the costs of their reproduction with the exchange value of their service. This results on the one hand from the low wage level and on the other hand from the high cost of living. Therefore, they are compelled to gain a selling advantage at any cost, even if it violates the limits of their integrity. It is not the lack of professional ethics, but it is the structural conditions that generate the risky practices and promote the exploitative structures. The women are dependent on this work and often cannot say "no" if they want to survive.

In "drug prostitution", the degree of alienation and dissolution of boundaries is obviously so great that it is no longer perceived as work by many sections of mainstream society and its scientific and political institutions, but only as a social and health policy or criminological problem. This section has shown, however, that the field of "drug prostitution" can be described without contradiction using terms from the critique of capitalism, and thus the concept of work would be an entry point for social work to fight effectively against the grievances and not against the individual life choices of the women.

4 "Drug Prostitution" Is Work: The Change of Perspec tive

The degree of exploitation and violence within "drug prostitution" and the individualization and desolidarization among the women working there is closely linked to the lack of recognition of their service as work. To counteract this problem, an awareness must be created both in society and among the women that their activity is work. Only when "drug prostitutes" have become drug-using sex workers through the adoption and acceptance of a positively connoted definition of work can new paths of empowerment be taken. This awareness must be integrated into the everyday counselling work of the social projects and implemented in their external presentation.

Sex work and not prostitution is the term that is suitable to adequately express the diversity and contradictoriness of labour and employment relations as well as the relations of domination, exploitation and violence in the field of sexual services. It is important to speak of work here as well, in order to make it clear that those working in this field have not yet been able to invoke minimum standards under labour law. State security systems hardly ever take effect (Schrader, 2013, pp. 88–93 ff.) and the drug-using sex workers find themselves in dependencies that often constitute a relationship of violence. The statements in the interviews I conducted prove that for most women sex work is a job that they do not do voluntarily, but is forced upon them by their life situation (Schrader, 2013, p. 40 ff.). At the same time, sex work is also normality and is described as a game with identities (Schrader, 2013, p. 390 ff.). A lack of a concept of work in the field of sex work contributes, among other things, to the fact that this service is devalued and demanded significantly below value. Therefore, it is necessary for an emancipative movement to proclaim such a concept (El-Nagashi, 2010, p. 17).

Based on Winker's design to understand work as a "purposeful and goal-directed social practice of human beings, which directly or indirectly serves to maintain one's own existence, the existence of other persons as well as following generations" (Winker, 2009, internal working paper), I propose the following conceptual definition for the sexual service of female drug users (Schrader, 2013, p. 404).

Sex work is a purpose-bound and goal-oriented self-technology of people, which directly or indirectly serves to maintain their own existence and that of other people. In this form of work, sex workers have a part of the means of production at their disposal, but they are always subject to the capitalist logic of exploitation and thus to exploitation. Sex work takes place in a space steeped in ideology and domination and always interacts with identity, structural relations of power and domination, and norms and values regardless of whether it is performed voluntarily, coerced, or as an insight into necessity. Drug use within sex work is not only a special and cost-intensive variant of consumption, as the term suggests, but it is also an important part of the reproduction of labour power in the special work situation with its hard working conditions.

This definition also conceals a political construction, which is justified for the time of political struggles for the participation in rights and their implementation. But this struggle should not stop there, since the field of sex work is far too heterogeneous for one definition to capture and take into account the rights of all those working there. An emancipative movement of individuals requires the partial and situational unification of different status groups that (can) have disparate ideals and goals.

The definition is also an analytical grid in order to be able to classify the work in its ambivalence of resistance and subjugation with the intersectional multi-level approach (Winker, 2012, p. 23). In this way, the work situation can be analysed and options for action can be worked out on the basis of interviews and in conversations with the women.

Based on a concept of work for sex workers,[2] social work can then return to political action for and with the "affected" in addition to individual counselling and survival assistance, in order to fight against sexualised violence and exploitation in sex work on the basis of long-term empowerment strategies, for example, for the establishment of qualification modules and their reliable financing (Schrader, 2013, p. 395 ff.).

If a society wants to successfully fight against sexualised violence and exploitation in the work environment, it is completely irrelevant whether we are talking about female doctors, soldiers, domestic workers or sex workers and the nature of their respective "addictions", the degree of injustice remains the same. Sexualized violence must be recognized as a society-wide phenomenon in all fields of work (Brenssell, 2012, p. 209). Only in this way can a solidary and thus powerful union emerge from the individual-situational affectedness.

References

Brenssell, A. (2012). Gesprächsangebot zur „alltäglichen Lebensführung“ – Kritische Psychologie trifft Kritische Sozialarbeit. In U. Eichinger (Hrsg.), *Soziale Arbeit* (S. 190–260). Argument.

Butler, J. (1993). Kontingente Grundlagen: Der Feminismus und die Frage der Postmoderne. In S. Benhabib, J. Butler, D. Cornell, & N. Fraser (Hrsg.), *Der Streit um Differenz* (S. 31–58). Fischer.

Egartner, E., & Holzbauer, S. (1994). *Ich hab's nur noch mit Gift geschafft*. Centaurus.

El-Nagashi, F.-A. (2010). Sexarbeit und Migration. *an.schläge, das feministische Monatsmagazin*, 17.

Foucault, M. (1995). *Der Wille zum Wissen. Sexualität und Wahrheit I*. Suhrkamp.

Gerlach, R. (1998). *Gibt es „schwerstabhängige“ Methadonpatienten oder werden sie „gemacht“?* Vortrag zur A.I.D.-Tagung: Mehr als abhängig? Versuche mit Methadon und Heroin 15./16.05. Berlin.

[2] I lean here on critical psychology, which forms concepts not about people but for them from the point of view of the subject (Brenssell, 2012; Holzkamp, 1985).

Gerstendörfer, M. (2007). *Der verlorene Kampf um die Wörter. Opferfeindliche Sprache bei sexueller Gewalt*. Junfermann.

Holzkamp, K. (1985). *Grundlegung der Psychologie*. Campus.

Kappeler, M. (1998). *Drogen und Kolonialismus. Zur Ideologiegeschichte des Drogenkonsums*. Verlag für Interkulturelle Kommunikation.

Kavemann, B., Rabe, H., & Fischer, C. (2007). *Vertiefung spezifischer Fragestellungen zu den Auswirkungen des Prostitutionsgesetzes: Ausstieg aus der Prostitution. Kriminalitätsbekämpfung und Prostitutionsgesetz*. Sozialwissenschaftliches FrauenForschungsInstitut Berlin. http://www.bmfsfj.de/RedaktionBMFSFJ/Abteilung4/Pdf-Anlagen/prostitutionsgesetz-gutachten-2,property=pdf,bereich=,sprache=de,rwb=true.pdf . Zugegriffen: 2. Jan. 2009.

Kerschl, A.-V. (2005). Beschaffungsprostitution und ihre Risiken. In M.-T. Wright (Hrsg.), *Prostitution, Prävention, und Gesundheitsförderung. Teil 2: Frauen* (S. 113–122). Deutsche AIDS-Hilfe e. V.

Leopold, B., & Grieger, K. (2004). Gewaltprävention durch Arbeit mit Minderjährigen in der Prostitution. *APuZ, B, 52–53*, 19–26.

Meletzky, F. (2012). *Tatort. Wegwerfmädchen*. Arbeitsgemeinschaft der öffentlich-rechtlichen Rundfunkanstalten der Bundesrepublik Deutschland (ARD). Anstalt des öffentlichen Rechts.

Meyer, C., Neumann, C., Schmid, F., Truckendanner, P., & Winter, S. (2013). Bordell Deutschland. Wie der Staat Frauenhandel und Prostitution fördert. *Der Spiegel, 22*, 56–65.

Mitrovic, E. (2004). Der gesellschaftliche Umgang mit Prostitution in Deutschland. *Die Wüste lebt*. Regenbogen. HWP 14. Februar.

Müller, S. (Ed.). (1986). *Verstehen oder kolonialisieren? Grundprobleme sozialpädagogischen Handelns und Forschens*. Kleine.

Schrader, K. (2013). *Drogenprostitution – Eine intersektionale Betrachtung zur Handlungsfähigkeit drogengebrauchender Sexarbeiterinnen*. transcript.

Ward, H. (2007). Marxismus versus Moralismus. Trend 7–8/07 onlinezeitung. http://www.trend.infopartisan.net/trd7807/t407807.html . Zugegriffen: 21. Sept. 2014.

Winker, G. (2009). *Internes Arbeitspapier zur Handlungsfähigkeit in der Kritischen Psychologie*. Vortrag 15. Januar, Hamburg.

Winker, G. (2012). Intersektionalität als Gesellschaftskritik. *Widersprüche, 32*(126), 13–26.

Zurhold, H., & Kuhn S. (2004). „Ich komme an keinem Stein mehr vorbei" – Crackkonsum von jungen Frauen in der Hamburger Drogenprostitutionsszene. In W. Schneider & R. Gerlach (Hrsg.), *Schriftenreihe: Studien zur qualitativen Drogenforschung* (S. 231–254). VWB.

Kathrin Schrader has worked for many years in various social work projects, including in the field of prostitution. Since 2014, she has been a professor of social work with a focus on people in precarious situations at the FH Frankfurt/Main. Previously, she was a research assistant in the work-gender-technology department at the TU Hamburg-Harburg.

Social Work in the Context of Prostitution: Professional Approaches in the Field of Tension Between Different Systems and Actors

Julia Wege

1 Introduction

A central task in social work is to understand, analyse and advise people in their living environment and their systems. In particular, social work focuses on those people in society who are marginalised, socially excluded or in need of advice and help in different contexts. Particularly severely disadvantaged target groups, who are least able to stand up for their rights and draw attention to their problem situation, are hardly heard in public and should be increasingly the focus of social work. Although social work has developed enormously in terms of professionalisation in recent years and is constantly working to help people in different situations of need, there are still areas which are insufficiently focused and not scientifically evaluated. Homeless people, asylum seekers, criminals, addicts or women or men who work in prostitution receive little attention in society or in research. These target groups do not exactly belong to the preferred clientele. However, it is precisely this group of people, who are often caught up in the help system for years, that constantly pushes social workers to their limits. The complex changes within society, the increase in the number of people living in poverty and the different groups of clients who are dependent on help are constantly presenting social work with new challenges (cf.

J. Wege (✉)
Mannheim, Germany
e-mail: julia.wege@rwu.de

M. Albert, J. Wege (eds.), *Social Work and Prostitution*,
https://doi.org/10.1007/978-3-658-37761-8_5

Wege, 2012). Concepts of action specifically geared to the problems and professionally and self-confidently acting social workers are necessary to be able to meet the requirements in practice. Only in this way can social work in difficult living environments or milieus be effective in the long term and also be regarded with esteem by politics and society from the point of view of prevention and cost savings.

This article examines social work in the context of female prostitution and describes the different systems and actors in the field. The attempt is made to show which difficult framework conditions prevail in the aforementioned living environment and how this affects the field of work of social workers. Ultimately, this raises the question of what competencies are required for social work to be professionally and sustainably successful as a human rights profession in the field of prostitution. The legalisation of prostitution in 2002 has not only changed the situation of women, but also the situation of social work within the field of action (cf. Reichert & Rossenbach, 2013, p. 6). It should be noted that social work can already demonstrate professional approaches to action in this area, but these must be further expanded and professionalised in the future.

Social work basically pursues the goal and has the high demand to offer high-quality counselling to the addressees. However, it is often influenced in its work or even put under pressure by external framework conditions such as case numbers to be provided, financial savings measures or a lack of acceptance by the public. Social workers distinguish themselves by their ability to respond empathically and flexibly to people in the shortest possible time, to recognise complex problem situations and to develop solution strategies. In this context, it is necessary to constantly adapt to new situations and to always keep the individual person with his or her individual problems in focus. For this purpose, social work has a wide range of different methods and techniques for conducting conversations and providing counselling from which it can draw. In the counselling process, it is not only important to take into account the environment in which the client moves, but, as Alice Salomon (cf. Galuske, 2007, p. 78) already addressed in the early days of professional social work, the interaction between the individual and his environment should be included in the assistance process. This goes hand in hand with analysing positive as well as negative influences and mobilising resources and abilities of the person seeking help. Another aspect is to take the perspective of the addressees of social work and to understand them in their living environment and to stand up for their rights and interests, regardless of whether it is a homeless person, a criminal or a prostitute.

Social work in the field of prostitution is subject to much more complex external conditions than, for example, youth social work. While this kind of social work is highly respected in society and has its raison d'être, social work in the field of prostitution is often questioned. Although there is a great deal of media and voyeuristic interest among the general public, who knows exactly what the actual reality of women's lives is like? The really important question is: Who has access to the women working there? It should be noted that prostitution is a milieu of its own, which usually takes place in a so-called parallel society. The topic of prostitution is often concealed and disregarded by society or politics. In addition, there are prejudices in society and not infrequently false opinions about the sex trade, which do not correspond to reality. Only very few people know about the actual causes and backgrounds. The fact that women from the prostitution milieu are in urgent need of help and support is often not understood by society. Due to this initial situation, it is difficult for the population to understand why women in prostitution in particular need counselling services. Prostitution is often associated with a criminal milieu and a social "grey zone" which is ethically and morally rejected. The milieu finds its acceptance and legitimation in the fact that the male sexual drive finds its satisfaction, which in itself triggers enormous curiosity and interest. In most cases, however, no further contact is desired. In cooperation with specialized authorities and social institutions, it is not uncommon in practice that the fear and lack of specialized knowledge about prostitution is very great and leads to a certain powerlessness and blockade. In counselling interviews, this insecurity manifests itself in an unprofessional or aggressive way of talking to the women. Negative statements or one's own opinions about prostitution in the presence of a client who, for example, is trying to get out with the last of her strength and resources, do not contribute to a positive new start without problems. When leaving prostitution, women usually want to leave their past behind and come to an end with the prostitution phase of their lives. Discriminatory or even "crude" statements in counselling interviews reflect the ambivalent and contradictory attitude of society. For the women, this is an additional reason to keep quiet about their past and to approach new encounters or relationships with great scepticism. It is not uncommon for them to avoid cultural or social participation in society in order to avoid possible negative confrontations and remarks.

When considering the extent to which cities regulate prostitution at the municipal level, it becomes clear that prostitution is located in a legally difficult grey area that can hardly be controlled and monitored. Complaints from residents, the increase and spread of prostitution establishments and street prostitution or a lack of influence on the accompanying phenomena of the milieu are frequently occurring problems that are often exacerbated by media reporting. In the attempt to find holistic solutions, strategies and action concepts of different cities are always compared in such discussions in order to be able to weigh up the pros and cons. It does not seem to happen often that scientific analyses are commissioned in order to develop conceptual proposals for solutions on the part of social work. However, this kind of approach by social work research is indispensable, since regulatory measures are not sufficient at this point to achieve a holistic and long-term improvement of the situation. Such corresponding scientifically founded investigations offer an optimal basis to conceive fitting concepts in the sense of a positive and also holistic urban development. Quickly developed and not completely thought-out concepts, which arise from uncertainty and increasing pressure from politics and the media, must be viewed with scepticism. In this case, failure is possibly already pre-programmed. The milieu of prostitution cannot be compared with other professional fields of social work and therefore requires a longer planning period, which should not be underestimated. The lifeworld of women must always be of central importance in such concepts. The elementary question of the extent to which women can be reached and advised by social work services is enormously important. Pragmatic and feasible proposals for solutions only make sense if they are constantly reviewed and improved during the implementation phase.

Now the question arises in which systems social workers move in the field of prostitution and out of which necessity a closer look seems to make sense. In systemic thinking, systems are spoken of on a meta-level in that "...all persons, things and their interrelationships can be described as a system and these are in a relationship with each other" (Lindemann, 2008, p. 6). These systems have a basis to act or interact together in a certain way, even though each individual person takes his or her own particular view of the system in question. The principle is that systems can change at any time. A change can already be set in motion as soon as a system member behaves differently than usual. Therefore, there is no final system, since systems are in motion and each system has its own development and fluid function. Each system also has its system boundary to the so-called system environment, which does not belong to the system. According to Lindemann (2008, p. 6), systems are "observer-dependent descriptions" because perception, attitude and knowledge are included in the system description. Such a description of the system already excludes other systems. There are two ways in which systems are distinguished in systemic thinking: the trivial system and the non-trivial system. If we assume that systems always run in the same pattern according to the cause-effect principle, there is a danger that we will overlook so-called blind spots in this trivial system approach. Trivial systems can be analysed and future predictions can be derived from them. From a systemic point of view, social work in the context of prostitution thus operates in different systems and their interactions, which will be outlined in more detail below. It should be critically noted that the action of social work is often a reaction to corresponding grievances and then only becomes active and has its concrete mission when the need is greatest. It would be desirable if social work could already act beforehand or preventively and have a positive effect on the corresponding systems and their problems.

2 The System of the Public or Society

In this system, certain values or positions about prostitution are represented. However, these can be completely controversial and are usually influenced by the one-sided or negative reporting of the media. Women who work in prostitution belong to a marginalized group that is constantly exposed to stigmatization and discrimination. On the one hand, the image of the self-confident prostitute or sex worker prevails in society, who sees the activity as part of a self-determined sexuality. This view objectifies the milieu of prostitution and justifies the raison d'être of prostitutes for acting out the male sexual urge. Without any scientific basis, prostitution is understood as a preventive social measure in that it reduces sexually violent crimes against women and stabilizes partnerships or marriages through prostitution. Prostitution can also serve as a social and stabilizing compensatory function by serving those clients who have sexual difficulties, are disadvantaged or are inhibited, for example, due to physical disabilities (cf. Brüker, 2011, p. 25). For some years now, prostitution has been upgraded and its image changed with skilful marketing strategies. The aim here is to lower the inhibition threshold for the male population and to constantly attract new customers with promising offers and campaigns. Prostitution has long since ceased to be the business of "sex for money" per se, but is increasingly becoming a wellness and recreation area for stressed men from all walks of life. It can even be spoken of a new sector in the lifestyle area, where business successes are also celebrated without moral concerns, for example. Additional offers such as sauna, massage, upscale catering, discounted daily admission prices or annual passes for permanent guests contribute to basically disguise prostitution and facilitate access for men or especially for new customers. The uncomplicated entry into the milieu with high-level framework conditions and offers has led to prostitution becoming increasingly socially acceptable. In recent years, the infrastructure in the prostitution sector has expanded enormously, although the demand side continues to take place in the imperceptible area (cf. Gerheim, 2013, p. 40). While media coverage and discussions in society and politics are always oriented towards the plight of women, it can be seen on the other hand that the demand side continues to be silently overlooked. No man wants to be outed in the media as a john, the social exposure in public would be too great. This makes it all the more necessary to discuss values within society with the men who use sexual services. For they have direct access to the women and can see, if they want to admit it, whether a woman seems intimidated and possibly works under duress. If the demand side would deal with women from prostitution in a more appreciative and respectful way and would have the courage to report abuses, the working conditions and thus also the public recognition of women in our society would change significantly.

Another view of prostitution is that it is evaluated as exploitation and a violation of human rights. Here, women are seen as victims and as subjects or commodities in a male-dominated society: "In this context, prostitution represents an existential attack on women's right to sexual self-determination and degrades them to an object of exchange for male-sexual desire to submit" (Gerheim, 2013, p. 41). In this case, society often reacts with great consternation and willingness to help by donating or volunteering to do good in order to alleviate the need. However, this view can also tend in a completely different direction, in that the women are devalued, insulted and the issue is played down or, at best, completely ignored. In this view, it is not immediately apparent and recognizable why these women are in need of help. At this point, there is no differentiated and in-depth analysis of why women find themselves in such complex predicaments, as the assumption is made that women themselves are to blame for their predicament. A further examination of the topic is not desired or is usually completely blanked out and defamed. Misogynistic or racist remarks are additionally voiced to reinforce one's own position. In social work reality, women in prostitution are on the lowest fringe of society and even there still represent an exceptional group.

While the discourse on the monogamy requirement evaluates prostitution from a religious-moral perspective in terms of practices immoral and as adulterous, a further consideration of the "inner male competitive and distinctive struggles" is about the inability of the john to have to live out his sexuality with prostitutes: "The demand for prostitution is thereby devalued as an inferior sexual practice and marked as a subjective failure of sexual self-presentation" (ibid., p. 40). The john is not able to enter into or establish sexual relations under normal circumstances and on a mutually voluntary basis without money. Of course, there are mixed forms in the described patterns of public perception, with polarising, voyeuristic opinions often being found in practice. For counselling centres, due to this difficult handling, a clear public positioning, differentiated and regular reporting, the holding of serious events and the publication of annual reports are therefore of particular importance. In this context, it becomes clear that social work also takes on a preventive mission and, through educational campaigns in various bodies or other social contexts, points out the issue of prostitution in a factual and scientifically sound manner and puts it up for discussion. In a certain sense, social work must use professional concepts to bring the complex topic of prostitution from the "dark to the light". Only in this way can a socially important acceptance and sensitization be achieved, which is considered urgently necessary. It should be noted that so far various scientific disciplines such as psychology, sociology or law have approached the research topic of prostitution. However, it is noticeable that social work is the only science of action that is actually active in the red-light scene not only through research, but also through concrete counselling offers, projects and interventions. The logical consequence of this is that social work has the legitimacy to express itself critically as a profession based on professional ethics and human rights, and to develop recommendations for action in order to be heard in society and politics.

3 The System of Cooperation Partners and the Professional Public in the Municipal Context

Although social work refers to people and their problems, it must also always focus on the social space in which the target groups live. When it comes to the socially critical issue of prostitution, social work has to deal intensively with politics and the administration and act professionally in the system. The system of the professional public includes all important cooperation partners such as municipal authorities, social institutions, representatives of politics and the police and those persons who come into contact with the topic in any way due to their professional function. In every city, this system and its patterns of relationships and networks can be very different and has its own dynamics, the structure of which must be analysed and strategically thought through by social work. If a counselling centre for women in prostitution has already existed for decades, it may have already fought for its right to exist and may be able to profit from a respectful reputation and image in the professional public. Statements or demands towards cost units or politics are not only noticed but also taken seriously and thus increase the chance of actually being implemented. It is recommended that the social workers of the counselling centres contribute their expertise to various working groups and participate in current socio-political discourses. All public sectors can be of importance where their competences are required and in demand. This leads to a higher acceptance in the professional public and the points of view become more transparent and comprehensible. Social work has to fight for its firm place in the social help system, in which it wants to act and cooperate, in a certain way. It is not uncommon for employees of youth welfare offices, pregnancy or migration counselling centres to turn to their colleagues in the counselling centres with great uncertainty and to be dependent on professional advice. In this context, a cooperative collaboration, for example in the form of a goal-oriented and custom-fit "tandem" counselling, proves to be optimal, in which specialists coordinate with each other and complement each other optimally.

Counselling centres that are newly established or have not existed for long, however, have to work hard in many respects to gain a professional reputation. The funding agencies, politicians and the general public wait and watch critically to see how the project will start, what case numbers will be reached and whether the counselling services will be accepted by the clients at all. In the start-up phase of new projects, networking and maintaining contacts must have top priority. The help system of the city or the social areas must be analysed and recorded in detail. It is necessary to know which problem situations are present in the social area and which other social (marginal) groups are in particularly precarious living situations, especially since there is often a close connection here. In addition, it is of central importance to know about the different social organisations, counselling services and social support systems or projects. It has to be considered concretely where future common interfaces will arise and where cooperations have to be strived for or established. This process can take quite some time and usually costs a lot of energy and persuasion. The exchange with the police and politicians should not be underestimated. Information from the police is necessary and plays a not insignificant role in locating places of prostitution, addressing possible problems and discussing cooperation with the various departments. As strange as it may seem, the police in particular, albeit with a very specific remit, are very close to the realities of the milieu alongside social work. In addition, the police can intervene as a law enforcement agency and ensure victim protection for the women in individual cases, while the counselling centre for women in prostitution can provide psychosocial care and support (cf. Tanis, 2013, p. 47). The exchange with politics is relevant in order to point out the specific problem and to elicit possible funding possibilities. In particular, the political parties should be taken into account comprehensively and, at best, all important actors should be invited to a "round table on prostitution". As an example, the "Round Table on Prostitution" of the state of North Rhine-Westphalia should be mentioned here, which among other things has set itself the goal of improving the situation of women in prostitution. The interdisciplinary "Round Table on Prostitution" of the city of Mannheim, for example, meets quarterly to provide technical and conceptual support for the establishment of the "Amalie" counselling centre. Scientists and experts are continuously invited to benefit from the findings and experiences of other cities. At this point, social work must present itself self-confidently as an expert for the social and, for example, independently steer and take charge of the organisation, implementation and moderation of such committees. There is often a danger of becoming a passive participant itself and being dominated by other authorities or administrative apparatuses, instead of taking responsibility and steering such processes independently as an active actor. Prostitution should not only be considered from a regulatory, police or legal perspective, but also from a social work perspective, which is based on theories and ethical values of their profession.

4 The System of the Prostitution Milieu

The prostitution milieu belongs to a system that has its own specific dynamics and functions according to its own rules and structures (cf. Wege, 2010). These rules are clearly defined by the various "actors" in the milieu. The red light, which is actually shielded from the outside world, orients itself to social changes and is constantly evolving. Whereas in the past women from Southeast Asian or African countries or German women worked in the sex trade to a large extent, today it is mainly migrant women from Eastern Bloc countries such as Bulgaria or Romania. These processes of change are the result of increasing globalisation and the associated opening of borders and, as a consequence, are subject to clear economic market mechanisms and power interests. In a certain sense, we can speak of a constantly evolving and adaptive system, which constantly adapts to the social and also legal framework conditions. According to Girtler (2004, p. 274), the milieu belongs to a subculture, which is even oriented towards the upper class. It has its own everyday knowledge, code of honour and language. The actors pursue identical goals, such as wealth or corresponding status symbols, which are present in the upper social class. In particular, this includes leading a luxurious life and being able to show off corresponding goods. However, Girtler emphasizes the striking difference in the achievement of goals. Actors of the milieu use other or criminal means and ways to achieve their goals, as is common in the bourgeois social class (ibid.).

Each city has its own historical development with regard to the red-light milieu, which must be taken into account and included in the entire analysis. Milieus establish themselves socio-spatially where hardly any resistance prevails and their existence is accepted. These can be secluded streets in residential areas, in the consumer-oriented inner city area, or even commercial areas outside of urban spaces. Often, however, they are streets in problem-ridden neighbourhoods with a high occupancy rate, characterised by residents of different nationalities. Unemployment, immigration problems, impoverishment and poor building fabric in these residential areas often go hand in hand with structural or social neglect. In such residential areas with an increased need for development, the red light scene often establishes itself and is accepted by the residents in different ways. Often, due to social and public pressure, the red-light area is downright displaced to other parts of the city and only tolerated there (cf. Reichert & Rossenbach, 2013, p. 3). In this context, regulatory measures play a decisive role for urban or red-light development in terms of the extent to which the milieu expands and spreads into neighbouring residential areas. In middle-class or more affluent neighbourhoods, on the other hand, red-light districts in public spaces would be unthinkable or unenforceable. The resistance of the better-off residents would be too great, who would fear a loss of image and, as a consequence, a loss of economic value. In this context, prostitution is subject to the power of money in many facets and it is completely neglected that it is ultimately about poverty development and its effects on individual life plans. At this point, the contradictory attitude of the state is clearly visible. On the one hand, in 2002 the legalisation of prostitution was described in a political consensus as a milestone for a liberal attitude and sex work was recognised as "normal" gainful employment. On the other hand, women in prostitution are pushed out of the publicly perceivable space and merely tolerated in remote districts. The liberal attitude and approach towards prostitution requires, in a real sense, that space and room be created for this trade. Here it must always be taken into account that there are great differences in large cities or rural areas – there are simply worlds between a Lake Constance region and a large city like Mannheim.

The ambivalent attitude between toleration and great desirability on the male demand side is reflected in many facets. For the practicing women in the prostitution sector, this means being constantly observed or controlled by regulatory authorities, unless they have their permanent establishment where they can show a permanent job. Terms such as pimp, landlord, john or trafficker appear in all contexts and different combinations in this system. The lack of knowledge about this area increases the scope for subjective patterns of interpretation. The lack of scientifically sound data and meaningful knowledge about the trade, the women as providers and the demand side of the men favours that prostitution is talked about more in myths and half-truths. As the milieu has taken on an even greater dimension due to poverty prostitution and far-reaching problems have arisen, exploitative structures and business models have emerged that contradict the liberal attitude of the Red-Green government of the time. Over a long period of time, the system was left to its own devices, with the result that the complex problems have now been recognised by politicians and new laws and regulations are being demanded.

Now the question arises what role social work plays in this system. Social workers establish important contacts in the milieu in the form of outreach work in order to introduce themselves and the activities of the counselling centre and to make them known. In doing so, they have to adapt to the rules and structures they find.

Talks and contacts with landlords, housekeepers or clients are just as much part of the methodological concept as contacting the women who are the actual target group. When social workers enter a brothel to make contact with the women, it is indispensable to introduce oneself briefly at the door or reception in order to reduce the fears or critical apprehensions of the counterpart. It is of particular relevance to know the right contact persons in the business, who have an important function or act as multipliers and door openers. It is not uncommon for various local contact persons to approach the counselling centres in emergency situations and ask for concrete help and support for individual women before an ambulance or the police are called. The different interests of such contact persons and the counselling centres must of course be clearly distinguished. While consensual and benevolent co-operation is of central importance for social workers, it is also associated with contradictions, ambiguities and possibly dangers. A negative attitude would lead to counselling centres being denied entry into the milieu in which they have to establish the important contacts with women. It is not uncommon for clients to turn to counselling centres for advice on how they can best support women in emergency situations. In this regard, the communication and interaction is rather simple and uncomplicated. It can also happen that women are brought directly to the counselling centre by them and are also picked up from there after the counselling appointment. They seem to be downright caring clients who, according to the clients, are married and have children themselves, but are highly committed to the women's needs. Systemically, the attribution of the role of co-therapist could apply. Through this form of support, clients may be able to calm their own moral concerns and guilty consciences. At this point, however, the role of the john should not only be portrayed as caring, as there are definitely other behavioural patterns that have extremely negative effects on the women concerned.

Knowledge about the milieu should be consolidated through participation in professional discourses, discussions with experts or exchange with colleagues from other counselling centres. Those who go into the milieu without knowledge have to expect rejection up to house bans or other reactions. Social workers must therefore have a sound knowledge of this system or this subculture in order to be able to successfully orient their solution strategies accordingly. Only in this way can professionals meet with acceptance in the milieu and among the women. In this respect, it requires a special competence to be able to act professionally and self-confidently on the part of social work.

5 The (Poverty) System of the Women Concerned

Women who work in prostitution operate in a system that is inconceivable for normal bourgeois society. Terms such as "prostitute" or "sex worker" or "victim of human trafficking" do not contribute to the fact that women are seen value-free as individuals. The terms used mean that women have to live with negative stigmatisation. Due to the lack of distinction between the individual fields of work and poorly targeted discussions, in which it is often not clear whether one is talking about a poverty prostitute or a self-determined sex worker, such discussions lead to confusing misunderstandings. It is not uncommon for experts in socio-political and professional bodies to lapse into a discussion of values or, in some cases, to become hostile to one another. A new definition of the term would be advantageous at this point, as it could clear up ambiguities in the entire discussion debate in advance.

Far too little is known about the system of women's lives. Mostly, women choose the path into prostitution in order to gain quick income and practice prostitution either short-term, permanently, as a side job or full-time. "It is the individual social constellations during certain ages that lead women to become prostitutes" (Girtler, 2004, p. 33). Entry is usually through acquaintances or friends who have already been able to earn a lot of money in prostitution. If the poverty prostitution of Bulgarian and Romanian women in particular is taken into consideration, then it becomes apparent that many women come to Germany because of the lack of prospects in their home country and in the hope of a better life. These high expectations and wishes often do not correspond to reality. In most cases, these migrant women are under great pressure to earn money quickly with all the means at their disposal. The aim is to pay the upcoming and running fixed costs such as the daily rent and taxes and to transfer the rest of the income to the home country so that their own children or family can live on it in the country of origin. Although the women have left their homeland behind and offer their bodies for sexual services in a foreign country through prostitution, they are thus the main breadwinners of the family of origin. They cannot afford to show weakness, to be sick or to live their lives according to their own wishes and ideas. Functioning in the system of prostitution, where money can be earned but often not spent in a self-determined way, shows the whole dilemma of women. For these reasons, prostitution has taken on a new dimension, which has intensified due to the EU's eastward expansion in 2007. With the opening of the borders within the EU, policy-makers were hoping for economic goals and gave little thought to the social effects and the influx of poor migrants. The lack of intervention by the regulatory authorities has allowed structures and illegal business models to emerge that are often incompatible with the values and also human rights of our society. Systems have emerged that exploit people in poverty in the form of low or underpaid hourly wages, promises of work and pretences. In urban poverty neighbourhoods, squalid apartments without electricity or sanitation are rented out to destitute or homeless migrant families at horrendous prices. Without language skills, education and health insurance, these people are forced to accept all offers of work in order to provide for their families. Immigrants who have already migrated take advantage of this plight and try to make a profit from this system by offering supposed help. In this context, prostitution is a partial aspect of the entire immigration system. For women who have chosen the path of prostitution voluntarily or involuntarily out of a situation of need, it can be difficult to leave this sector on their own and to reorient themselves professionally. If a large financial allowance is demanded weekly in the home country and male family members or supposed friends enrich themselves additionally with the income, there is hardly any pocket money of their own left at the end of the month. This is often justified by the traffickers or pimps with the argument that they have established the contacts, provide a constant supply of men and have to pay off debts. An exit, which is presented by society or politics as a quickly realisable alternative, is hardly possible under these conditions. The overall poverty situation and precarious living conditions force the women to remain completely in the milieu. The lack of perspective is responsible for the women having to sell their bodies more and more below value. Stress, fatigue, depression are the fatal consequences and thus often topics in counselling sessions.

At this point, social work merely assumes the function of psychosocial support or humanitarian survival aid. It is hardly possible for the women to get out of this predicament. An own apartment with deposit requirements and necessary expenses for furniture cannot be financed because there are no assets. The lack of German language skills makes it difficult for the women to look for accommodation on their own. Vocational training or other gainful employment is de facto not possible. In addition to arranging a language course, support in finding accommodation and sensitive mediation with the landlord would therefore be a first important goal. In this context, most women are particularly afraid of the question "What was your last job?" or "Can you bring your salary slips?" If they were to give truthful information, the women would lose their apartments or jobs again in a very short time and would have to reckon with being subjected to future discrimination or unpleasant comments from colleagues. So there is only one thing left to do: to keep silent about their previous life. Due to social stigmatisation, they are permanently forced to lead a double life. When women try to build a new life and make new contacts after a successful exit, they often have the added difficulty of being unsure of how to approach people and build new relationships. There is often competition among the women in the milieu, as it is a question of daily survival and who attracts the most clients. Friendly contacts are often very superficial and short-lived. Aspects of collegial solidarity are hardly found in these social milieus. The attractiveness of the women, their age, the prices they charge and the clothes they wear are factors with which they compare themselves among themselves. Their origin also plays an important role. In this context, racist statements and devaluations of other ethnic minorities also occur (cf. Le Breton, 2011, p. 175).

A high number of the women have their own children, whom they either have cared for by relatives in their home country or who live with them themselves. However, this can often be used as a means of pressure in the form of intermediaries forcing the women to transfer money regularly so that they are allowed contact with their children. Most women conceal from their children where they work and do not wish their own children to know one day. The division is also reflected in the fear that, in the worst case, their own children will one day work in prostitution themselves. It is striking that women in prostitution show a high degree of responsibility and caring towards their children. The existence of the children often seems to be the last ray of hope for the women to survive in the cycle of prostitution. A major difficulty, however, is when the children grow older and know about their mother's activity. Conflicts often arise here that are far-reaching, especially for the children. "If society were to accept prostitution as a normal profession, prostituted mothers would certainly have less difficulty in educating their children about the exact nature of their profession - at least after a certain age" (Campagna, 2005, p. 273).

6 Social Work Competences in the Professional Field of Prostitution

According to the competence model of Gromann (2010, p. 11) social work professionals have on the one hand the task to improve the living conditions in the client system and on the other hand to act in a complex social service system. This results in three areas of competence: self-competence, case competence and system competence. While case competence focuses on the contents, problems and resources of the case, system competence focuses on all information and network partners in the service system that are important in order to be able to act professionally. Self-competence refers to the social worker with his or her attitude, view, stance, competence and motivation. While the case and system competence have already been explained, the so-called self-competence, which social workers must have for the professional field, will be examined below.

According to Thiersch, lifeworld orientation is an important concept in counselling (see Grunwald & Thiersch, 2011, p. 854). It is indispensable to analyse the complex field of prostitution and the lifeworld of women across systems and to understand the interrelationships as a whole. Thiersch focuses on the dimensions of time, space, social relations and the work of coping. The dimension of time means that the biographical trajectories of the addressees must be dealt with respectfully, individual phases of life must be placed in context and considered in connection with the future. Another aspect is space, which plays an important role in lifeworld-oriented action. Social work offers for women who, for example, work on a remote street promenade in an industrial area, must be designed in such a way that they are also accepted and can have a lasting effect. A socio-spatial analysis and the knowledge about the milieu give social workers important hints on how to orientate their work and thus underline the aspect of professionalism. Women who work in prostitution often do not draw a line between their working and living space and often have negative experiences in this regard. However, a living space should also offer security and reliability. The aim of social work should therefore be to change this space for the women concerned so that they feel comfortable in their space. The last dimension is that of social relations. Social work addressees live in complex, often very stressful structures and need professional help to build and shape a new and reliable environment. In lifeworld-oriented work, coping work aims to develop strategies so that clients gain security in their way of life and also acquire the competence to cope with life in a balanced way. The lifeworld of women is differentiated by various factors, which can have either a positive or negative effect on the biography of women. The biographical life course plays an important role. The following questions are of central importance here:

- In which family circumstances did the addressee grow up?
- What values and norms were taught?
- To what extent was autonomy and education emphasized?
- What life circumstances led women to enter prostitution?
- What subjective life plans, desires and feelings are hidden behind women who also have a "name" or a biography?
- In what different systems and also roles do women act?
- Which systems or roles have a positive effect and which have a rather inhibiting effect on the biography and current life situation of women?
- What form does resource-oriented counselling take and what skills do the women develop through their work?

- How does sexuality as a basic need affect the living environment, language and sensations of the women concerned?
- How does the issue of women's sexuality affect the subjects of professional social work?
- How can these aspects be taken into account in counselling and included in a resource-oriented way?
- How does social work have to start in order to really reach women?
- How can all systems involved maintain a positive attitude in the tense relationship between sexuality and power or the complex relationship between man and woman?

These questions are helpful in the context of the method of individual case help for an activating anamnesis, the course of counselling and a critical self-reflection. In everyday counselling, social workers are always confronted with the complex life-world of their clients. In the counselling context of prostitution, special skills such as empathy, active listening and a non-judgemental attitude are necessary (cf. Albert & Wege, 2011, p. 14). A central role is the relationship between counsellor and client, which must be established with great care. Even minor errors in counselling or incorrect or judgmental or derogatory statements to the client can lead to disappointment and even to refusal or non-use of further assistance. Good quality counselling can only be achieved through conversations that are based on an equal footing and built on a foundation of mutual trust. These women sometimes need years of continuity in the relationship because they have not experienced this human quality in their biography. The problems that the women bring to the counselling and expect concrete help in this regard are quite complex, but also show a very high level of intimacy. The topics range, for example, from health problems in the gynaecological field to borderline violations, wishes to leave the sex trade and sexuality. However, they can also concern quite banal everyday questions such as "How and where do I open a bank account?" or "Where can I register my daughter in the next kindergarten?". On the part of the women, a very intensive closeness is demanded due to these specific topics, which sometimes also has aspects of a friendly relationship. A healthy balance between closeness and distance must be professionally reflected and controlled by the social worker. Experience in counselling women in prostitution shows that at the beginning of the counselling sessions the social worker is meticulously scrutinised and viewed with suspicion. However, once this initial phase of scrutiny is over, there is a very great need for closeness on the part of the women and a particularly great openness. They quickly talk about personal issues and the social worker often takes on the role of a good "friend" or a "mother", which they have never had in their lives. Here a very emotional and vulnerable side of the women is revealed. The need for attention, love and security is pronounced to a high degree. However, there is also another side that can come out at any time. Women in prostitution are used to having to assert themselves and negotiate. They are very present, at least outwardly and can appear with a dominant attitude of entitlement. If they are not immediately given the attention they desire, they can also become aggressive, demanding and transgressive. At this point it becomes clear in which highly stressed milieu the women live, which communication patterns they are used to and that one has to assert oneself in the scene with the elbow. Statements such as "Unfortunately, I don't have time to have a counselling talk with you until tomorrow" can already be interpreted as rejections, so that the woman hangs up angrily on the phone and does not get in touch for a week. In this situation, the disappointment is great because they were used to social workers helping them immediately with their questions and problems. If this help does not take place according to their expectations, they cannot cope with this situation and become rude, for example, and loudly express their disappointment. Social work should always be aware that in this field life dramas, psychological crashes and disappointments are part of it. Again, there are also those women who, with their reserved and cautious behaviour, tend to find it difficult to accept help and show great gratitude. The tasks and limits of counselling should therefore be clearly defined and expressed in advance during the first meeting. The danger can quickly arise that clients have expectations and wishes of the counsellors that cannot be fulfilled. They quickly transfer the entire responsibility to the counsellor with the aim of completely reorganising their lives, which may already have been beset with problems for several years. However, these excessive expectations usually cannot be fulfilled and require a realistic perspective.

Social work has its own methods, theories and concepts to ensure professional work. However, in the context of prostitution it is particularly important not to proceed in a stereotypical way, but to orient the counselling individually towards the women. The methodological concepts merely provide the open plan to act in a goal-oriented manner. The methodological competence, the individual counselling competence and the social worker's own personality should complement each other and be in constant positive interaction. Due to the complex issues and multiple problem situations of women, it is advantageous to have sound knowledge and skills regarding a large number of methods and interview techniques (cf. Stimmer, 2012, p. 25). Depending on the case, homelessness, pregnancy, migration, health, experiences of violence and trauma, and over-indebtedness, for example, can be central topics of counselling at the same time. This constellation of problem situations is the rule rather than the exception in counselling practice.

7 Burdens of Professional Counselling for the Social Workers

Prostitution is a multi-layered topic, which is afflicted with very large social taboos and takes place far away from the "normal" bourgeois-conservative society. This topic requires professional social work to question all ethical and moral values and opinions. The professional deficiencies in social work education are obvious: neither are there specific textbooks on the part of social work nor corresponding courses in the curriculum. Those who work in this field have to grow into the professional role themselves and build up a good foundation on the basis of theoretical knowledge in order to gain a professional identity. Methodological competence in individual case assistance, case management, community work, street work, group work and various discussion methods as well as specialist knowledge of (social) policy at local, state and federal level are absolutely essential. Networking, self-confidence and the ability to grasp systems quickly and to act professionally within them are further central aspects. While a social worker in debt counselling, for example, finds clear structures, administrative procedures, legal foundations and similar client systems, social workers in the field of prostitution often move in a legal, social and political grey area. In many places, structures have to be built up with a lot of energy and persuasion. It is not uncommon for professionals from other disciplines to prefer to refuse or even unconsciously "sabotage" support due to a lack of specialist knowledge, powerlessness and possibly also fears of acting wrongly (cf. Albert & Wege, 2011, p. 17).

The lack of willingness to cooperate with this clientele also reflects the ambivalent attitude of society here. Even if the legal situation is clearly defined and there is a legal entitlement, unclear reasons are verbally expressed, possibly underpinning rejections. Disparaging or diffuse opinions about the milieu are sometimes even included in counselling talks. Even when accompanied by social workers or volunteers, the women are not spared such statements in counselling talks at public offices, authorities or social counselling centres. Training for cooperation partners and other authorities in the form of sensitisation measures in counselling talks would be necessary and useful at this point. In the field of prostitution, social work often assumes a function to mediate and educate between society, women and the authorities (cf. Howe, 2012, p. 42). In addition, it takes on the function of lobbying for this marginalised group and uses the media to draw attention to the difficult living and working conditions of the women.

Counselling takes place in different places and requires professional flexibility on the part of social work. There are women who want anonymous counselling in a place of their own choosing, such as a café or a nearby park. However, counselling can also take place on the street during outreach work, in a brothel, on the phone or via online counselling. The frequently changing counselling locations and the lack of language skills on the part of the women require a high degree of flexibility, but also creativity and spontaneity. If there is already a lasting relationship of trust, the talks can also take place at the women's homes. These have a similar character to a home visit, which is known from outpatient assisted living. In the field of prostitution, it quickly becomes apparent that there are different access points and places where counselling is offered. The referral context, i.e. the context through which the women find access to the counselling centre, varies. The majority of women come into contact through street work activities and other colleagues or friends. In addition, there are inquiries from cooperation partners such as doctors, police, clinics and other counselling centres, for example from the area of migration or pregnancy counselling. Not to be underestimated are the approaches via media such as the Internet or press publications. The issues that women bring to counselling are usually very complex and interwoven. Often the women seek help when the need is greatest and there is acute danger.

However, the question also arises as to what stresses can arise for professionals in this field of professional practice. Such stress potentials often seem to be lost in the discussion, although they do play an important role. Professional work in this stigmatized field constantly requires that one's own actions be stimulated for reflection. Especially in highly stressed areas of work, it is important to be aware of the thematic field in which one is acting and what consequences can arise for one's professional actions as well as for one's private life. Soberly considered, counsellors in the field of prostitution are constantly or daily confronted with negative burdens, helplessness and fears. Stigmatisation and discrimination are a constant part of the reality of women's lives; in this respect, conversations in counselling practice often also have therapeutic aspects. Experience has shown that the need for discussion on the part of the women is usually very great. Social workers are required to conduct these conversations with a high degree of empathy and should, based on Rogers' client-centred conversation, verbally reflect the feelings and moods (cf. Galuske, 2007, p. 176 f.). In doing so, the emotional states of the women must be empathised with in order to be able to find the right words. This can also lead to "co-suffering", which in itself makes sense, especially since it also involves unconscious processes in the course of counselling. However, scientific research has proven that "...compassion activates the pain centre in the brain, which is responsible for processing one's own physical pain" (Hoffmann & Hofmann, 2008, p. 3). This means that the emotions are similarly empathized with as if they were one's own. The more intense the counselling and the re-imagining or reliving of various life crises, the stronger the impact of the emotions on the counsellors themselves. Managing one's own energy and switching off or at least partially drawing boundaries between one's professional and private life are key points that should be constantly kept in mind. One's own mental and physical health and the so-called psycho-hygiene or self-care should be taken into account. Empirical findings clearly show that counsellors with a high level of personal satisfaction and balanced well-being achieve better results and successful actions in counselling and therapy sessions (op. cit., p. 10). The lack of well-being or negative thoughts can thus also negatively influence counselling sessions. These are important scientific findings that should be taken into account in everyday counselling. Due to transference mechanisms in the relationship between counsellor and client, there is a risk that even secondary traumatisation can occur among professionals.

8 Future Need for Action in the Field of Prostitution and Social Work

Social work in the field of prostitution operates on different levels and is active in an area that is subject to a multitude of social taboos and values that do not correspond to the bourgeois norm. This field of work has developed enormously in the course of the original whore movement of the 1970s and 1980s to independent counselling centres, which are under different sponsors. However, one cannot yet speak of a complete professionalization. There are already professional approaches in the counselling landscape, but there is often a lack of personnel, funding and the corresponding structures to further advance professionalisation. Nationwide, the offers should be expanded considerably in order to enable a nationwide help network, especially since there is a high demand for social work. Especially in rural areas there are hardly any counselling centres. Social workers in this occupational field are characterised by a particularly high level of professionalism. On the one hand, methodological knowledge is required, as well as the ability to act professionally in the levels or systems mentioned. Flexibility, empathy and a strong personality, a well-founded professional ethic and an established professional identity form the basis. The difficult clientele and the accompanying phenomena of the milieu do not make it easy for social work and pose new challenges for all professionals. Many aspects and approaches are still uncharted territory for social work, which must first be professionalised. At this point, the thesis can be put forward that prostitution is one of the most difficult fields of social work due to its complexity and diffuseness. As already shown, all the topics of general social counselling such as pregnancy, homelessness, debts, addiction and violence etc. come together in this field. In addition, there is the ambivalent attitude of society, which can also be reflected in the cooperation with other authorities or counselling centres. The legal grey area contributes to the fact that often nobody feels responsible. For the reasons mentioned above, social workers tend to be lone warriors who have to assert themselves under difficult conditions from case to case.

In recent years, social work has increasingly developed from a "silent" to a "proud" profession in the public eye (cf. Albert, 2012, p. 289). It would be desirable for social work in the professional field of prostitution to position itself even more professionally and self-confidently and to join forces more strongly among themselves. Experience has shown that the work done in this field is very good, but often too "quiet"; it should take the right and dare to be "louder". Furthermore, social work in the field should be made more visible. Whoever as a social worker has understood how to act in the presented and sometimes very complex systems and to link specialist knowledge and competences with them in this context, has professionally internalised the principle of this work. The flexible and always reflected approach in the described field can also be understood as "creative professional competence" according to Albert (2012, p. 290). Finally, it must be demanded that action concepts on the part of social work research should be increasingly commissioned by financiers, welfare state agencies or politics and require a high level of support and appreciation. The funding of such projects must not be limited in time and set too low. Professional social work can only be effective if it is given the appropriate positive framework conditions. And only then can it really provide assistance that successfully improves the living conditions of women. To stand up for these women in particular is an obligation and thus ultimately a measure of the degree of humanity and the ethical values of our society.

References

Albert, M. (2012). Von der stillen zur stolzen Profession – Soziale Arbeit zwischen Funktionalität und kreativer Professionskompetenz. *Soziale Arbeit, 8*, 282–290.

Albert, M., & Wege, J. (2011). Soziale Arbeit und Prostitution. Handlungsbedarf und Entwicklungsmöglichkeiten in einem tabuisierten Berufsfeld. *Soziale Arbeit, 1*, 8–17.

Brüker, D. (2011). *Das „älteste" Gewerbe der Welt. Eine Untersuchung über die Lebenslage älterer Prostituierter. Dortmunder Beiträge zur Sozial- und Gesellschaftspolitik*. Lit.

Campagna, N. (2005). *Prostitution. Eine philosophische Untersuchung*. Parerga.

Galuske, M. (2007). *Methoden der Sozialen Arbeit. Eine Einführung* (7., ergänzte Aufl.). Juventa.

Gerheim, U. (2013). Motive der männlichen Nachfrage nach käuflichem Sex. *Aus Politik und Zeitgeschichte, 63*(9), 40–46.

Girtler, R. (2004). *Der Strich. Soziologie eines Milieus* (5. Aufl.). Lit.

Gromann, P. (2010). *Koordinierende Prozessbegleitung in der Sozialen Arbeit*. Reinhardt.

Grunwald, K., & Thiersch, H. (2011). Lebensweltorientierung. In H.-U. Otto & H. Thiersch (Hrsg.), *Handbuch Soziale Arbeit* (4., völlig neu bearbeitete Aufl., S. 854–863). Rheinhardt.

Hoffmann, N., & Hofmann, B. (2008). *Selbstfürsorge für Therapeuten und Berater*. Beltz.

Howe, C. (2012). Struktureller Wandel in der Prostitution. Zwischen Hurenbewegung und Sozialer Arbeit. *Standpunkt Sozial, 3*, 35–47.

Le Breton, M. (2011). *Sexarbeit als transnationale Zone der Prekarität. Migrierende Sexarbeiterinnen im Spannungsfeld von Gewalterfahrung und Handlungsoptionen*. VS Verlag für Sozialwissenschaften.

Lindemann, H. (2008). *Systemisch beobachten – lösungsorientiert handeln: Ein Lehr-, Lern- und Arbeitsbuch für die pädagogische und betriebliche Praxis*. Ökotopia.

Reichert, S., & Rossenbach, A. (2013). Wir wollen den Frauen Unterstützung geben. *Ein Gespräch. Aus Politik und Zeitgeschichte, 63*(9), 3–8.

Stimmer, F. (2012). *Grundlagen des Methodischen Handelns in der Sozialen Arbeit* (3., völlig neu überarbeitete und erweiterte Aufl.). Kohlhammer.

Tanis, N. (2013). Unterstützung für Betroffene des Menschenhandels. In C. Gunderson, A. Müller, & G. Teichert (Hrsg.), *Zwangsprostitution und Menschenhandel*. Lit.

Wege, J. (2010). *Der sozialarbeiterische Handlungsbedarf im Problembereich „Prostitution" am Beispiel der Stadt Mannheim*. Unveröffentlichte Master-Thesis, SRH Hochschule Heidelberg.

Wege, J. (2012). Wohnungslosigkeit – multiple Problemlagen, neue Zielgruppen und Versorgungslücken. *Soziale Arbeit, 11*, 408–414.

Julia Wege Social Worker (M.A.), is head of the counselling centre "Amalie" – for women in prostitution in Mannheim. Since 2009 she has been a lecturer at the SRH Hochschule Heidelberg for individual case assistance, community work and trauma work. Her research and publications focus on prostitution, homelessness and community work.

Outreach Social Work in the Field of Health Services for Sex Workers

Elfriede Steffan and Tzvetina Arsova Netzelmann

1 Introduction

Prostitution scenes are per se fragile entities that change strongly and rapidly depending on political, legal, communal and social conditions. Social and health services can hardly keep up with these developments. It is very likely that their service profiles will always lag behind the demand. A successful implementation of social and health services for sex workers depends on reaching them at all. This is a great challenge due to the framework conditions described above and the still prevailing social climate of stigmatisation of prostitution. Outreach work plays an important role in this context. With the introduction of the Infection Protection Act (Infektionsschutzgesetzes - IfSG) in 2001, this task is even specifically mentioned in § 19 IfSG, which describes the tasks of the public health service for the prevention and diagnosis of STI:

> *"The public health department shall provide, or ensure in cooperation with other medical facilities, counselling and screening regarding sexually transmitted diseases and tuberculosis. These shall and may be offered on an outreach basis to persons whose circumstances place themselves or others at increased risk of infection."*[1]

[1] Quoted from Steffan et al. (2003, p. 3).

E. Steffan (✉) · T. A. Netzelmann
Berlin, Germany
e-mail: e.steffan@spi-research.de; arsova-netzelmann@spi-research.de

M. Albert, J. Wege (eds.), *Social Work and Prostitution*,
https://doi.org/10.1007/978-3-658-37761-8_6

The many supporters of this law were aware that its implementation would also require new methodological approaches to reach sex workers.

Since the 1980s, public health offices and specialised counselling centres have had a wide range of experience of which approaches methodologically enable adequate outreach of social and health services. The legal situation before the introduction of the IfSG included the possibility in the Sexually Transmitted Diseases Counselling Act (Geschlechtskrankenberatungsgesetz - GeschlKrg) to impose regular compulsory examinations for the four STDs[2] named in the Act on certain groups of persons with "frequently changing sexual intercourse" (so-called HWG persons). No groups were specifically named in the law, but this practice of compulsory examinations was carried out exclusively for female prostitutes.

Since the 1980s, however, more and more health authorities, especially in the north and centre of the country, have started to offer anonymous and free services for sex workers and to publicise them through outreach work. Since then, a variety of methods for reaching different target groups in the field of sex work have been applied and evaluated in Germany. Outreach work in particular has been and still is the focus of methodological discourse. Today, for example, the Deutsche AIDS-Hilfe (DAH) regularly offers further training on the topic of "street work". Voluntary and anonymous offers of STI[3] counselling and diagnostics, accompanied by outreach social work in the field and counselling in the facilities, have been identified and evaluated as the most successful interdisciplinary "method tandem" in the practice strengthened by model projects (Leopold et al., 1994; Gusy et al., 1994). At the end of the 1980s and in the early 1990s, this was a good and positive opportunity to develop and describe new methods in an effective and evidence-based way. However, from today's perspective, it must also be noted that these model projects were predominantly funded in the context of HIV/AIDS prevention, i.e. with a very specific and limited view of the sex work scene. Unfortunately, the limited view of the funding municipalities and states has largely remained until today. It is true that the spectrum of funding measures today includes not only AIDS and STI prevention but also violence prevention, help for trafficked persons and exit counselling, and there is also a broad spectrum of social work approaches in the field of sex work. What is still lacking, however, is a fundamental

[2] STD for Sexually Transmitted Diseases. Named in law: Gonorrhea, Syphilis, Venereal Lymphadenitis and Ulcus molle.

[3] STI: Sexually Transmittable Infections.

understanding among many political decision-makers of the need for and equipment of STI counselling centres in health offices and of specialist counselling centres. Nor is there any systematic support for quality assurance and further developments through model funding, for example. This applies above all to the public health service (ÖGD), which is the focus of this article, especially after the introduction of the IfSG in 2001. Despite the increasing discussion of the topic of sex work from around 2013, overall both the offers of the ÖGD for medical care and social work taking place in this context have been decreasing for years. However, there are exceptions: some creative, adequate and continuously developed offers, especially in metropolitan areas (examples: Cologne, Hamburg, Berlin, Bremen).

The thematic spectrum of outreach work in the field today ranges from professionalisation in sex work in the context of so-called indoor training (e.g. DAH) and peer-to-peer work (e.g. Hydra, specialist counselling centre for sex workers in Berlin) to prevention of STI in the context of street work (e.g. health offices, specialist counselling centres, AIDS help centres, hustler projects) to publicising the offer of medical care and social support from the health offices. Modern communication technology also requires new adaptations of the work to the circumstances. For example, apps are now used for virtual outreach work (e.g. specialist counselling centre for sex workers "Madonna", Bochum) (MGEPA, 2014, p. 23) and counselling services are offered on the Internet. Some training courses on participatory quality development approaches and their testing in the services of health offices and specialist counselling centres are also among the measures that help shape the profile and quality standards of outreach work in practice (BZgA, 2015).

In this article we have to limit ourselves to the overview and presentation, we want to present outreach work in the field of health services of the ÖGD and shed light on the effects of the cooperation with the medical services.

2 Social Work in the STI (Incl. AIDS) Counselling Centres in the ÖGD: Equivalent Methodological Approach or Supplementary Accessory to the Medical Services?

Since the introduction of the above-mentioned Infection Protection Act (Infektionsschutzgesetz, IfSG) in 2001, the focus of government measures on STI has been on prevention rather than epidemic control. In § 3 "Prevention through education", the IfSG emphasises the public responsibility for prevention. With regard to STI, neither epidemic hygiene measures such as compulsory examinations for certain groups are provided for, nor are named reporting obligations. Section 19, on the other hand, defines compulsory tasks for the public health authorities in the field of STI. The counselling centres for STI (incl. HIV) of the public health service are to reach in particular persons with sexual risk behaviour who either have no access at all or very difficult access to general health care. These include, for example, people without health insurance as well as people who would not visit established doctors for other reasons, for example because they fear discrimination and exclusion in standard medical care due to their occupation (e.g. prostitution) or their sexual orientation (e.g. homosexuality). In order to facilitate access for these persons, § 19 IfSG authorises the public health authorities to offer their services within the framework of STI consultation hours free of charge and anonymously and to link this with outreach work. For patients without health insurance, the STI consultation hour of a health office is one of the few places where they can make use of services of the health system, but these are severely limited. The range of medical care varies depending on the equipment of the health offices (Steffan et al., 2003; RKI, 2012). The neediness of patients can go far beyond the possibilities of an STI consultation and in individual cases make immense demands on the staff providing care. Medical consultation hours and social work act hand in hand, but are effective as independent working approaches. The outreach work in the scene conveys prevention messages, provides access to health, social and legal counselling in the public health department and other institutions, and makes the medical offer of STI counselling in the public health department known. Social work is crucial for the reach of the offers of the ÖGD in this field.

3 Principles and Quality Characteristics of Outreach Work

Outreach work in the field of sex work is one of the many forms of social work in public spaces. The professional demands on social work in the field are high; in addition, there is often also a high physical strain due to visiting e.g. remote street prostitution areas, working hours in the evenings and at night and adverse weather conditions. The general demands on the methods and quality of this form of social work apply in particular to the area of street prostitution, which is closely observed by the public, and will be briefly described here.

Social work in public space has different tasks. On the one hand, it is aimed at stabilising people in their life situation and "thus aims at their social inclusion or integration (access to the labour market, housing market, health system, education, etc.)" (Krisch & Stoik, 2011). It meets the persons to be sought out with unconditional appreciation and acceptance, empathy, authenticity and respect (Schmid & Keil, 2001). These necessary basic attitudes and attitudes of the person-centered approach (Rogers, 1951) "put the person in the centre and trust in his abilities and are sufficient conditions for personality development through the relationship from person to person" (Schmid & Keil, 2001). This is because outreach social work also aims to expand the options for action of the persons concerned. On the other hand, social work has to mediate between different interests and conflicts that arise from the use of public space by different groups. Disturbances caused by street prostitution often give rise to complaints from neighbours and traders (Howe, 2011). In the context of outreach work, there is thus also the responsibility to "publicize the interests of disadvantaged people, or to support their publication" (Krisch & Stoik, 2011).

Outreach work must therefore meet diverse and high standards. In the following, we will highlight some aspects that are specific to outreach work in the context of the STI counselling centres of the health authorities.

- **Access and places: In the beginning is the research**
  First of all, the places and locations of sex work must be known in order to be able to visit them. This is "easiest" in the case of street prostitution areas, and the address of brothels and clubs is often known. Apartments and offers that are exclusively advertised via internet profiles are more difficult to find. In addition, the fluctuation is high and the addresses change frequently. A compilation of known addresses with the help of other actors from the field and social work in the context of "mapping" is a good way to gain at least a partial overview of places, their accessibility and the existing infrastructure, not only at the beginning of outreach work (Steffan & Arsova Netzelmann, 2012; WHO, 2002).
- **Safety and professionalism: always in pairs**
  There are many reasons why outreach work should always be done in pairs. First of all, there is the self-protection, but also the possible exchange among each other. Appearing in pairs also reduces the danger of being mistaken for competitors or clients. If street workers are named as witnesses in criminal proceedings, it can be of great importance to have another witness for one's own impeccable behaviour. Since street workers in Germany do not have the right to refuse to testify in court, this is especially important (Stiftung SPI, 2009).

- **Access: occurrence on site**
  Sex workers have often had to experience exclusion and devaluation and often initially encounter people who approach them in the context of the setting with mistrust. In particular, people with a migration background, who may only be involved in sex work in Germany on a temporary basis, may have had bad experiences (or none at all) with representatives of social institutions and authorities in their home countries and are correspondingly reserved. In addition, there may be language problems if the social workers do not speak the sex worker's home language. It is therefore particularly important to communicate the mission or the offer clearly and unambiguously. For access to the brothel or club, the acceptance of the operator etc. is also necessary, without "permission" from this side, access is not possible.[4]
  The use of specially trained cultural mediators makes it possible to address people from different countries and with different cultural backgrounds. Overall, a lot of experience and tact is required on site. The sex workers' place of work is sometimes characterised by bustling activity. In such situations, contact is hardly possible. Here it is important to find out when a new visit would be more suitable.
- **Use of material (business cards, information brochures, etc.)**
  Building trust only happens through personal contact. Materials can provide information and support the contact, but a quick impersonal distribution of such materials can by no means be described as outreach work. However, the material can also develop its own "recognition value", namely if such materials have already been displayed in the club etc. and the sex worker approached can better locate the social worker. However, it is important to ensure that the material is simple and comprehensible and available in the most important languages of the sex workers on site.
- **Advertising of the own offer in the institution or in the office**
  Making counselling and medical services known in the office/institution is a core task of outreach work. This includes presenting the offer, explaining the conditions for its use (e.g. confidentiality, free of charge, what is offered, at what times and in what languages) and minimising fears of the facility. Freely after Billy Wilder, however, the principle "*Thou shalt not be bored*" must also be taken into account, especially with such basic offers. It is there-

[4] The question here is whether it is really right that the police, public order offices and tax authorities should be able to go in and out of brothels and clubs at any time, while social and health institutions have to go 'door to door'. There are different views on this, but I am of the opinion that cooperation with health and social institutions should be made obligatory.

fore important not to fall into a routine and to rattle off the same "slogan" at every visit. What is needed here is creativity, active listening, attention and engagement: what issues were raised last time? Is there anything in the media of particular interest? Do I have an interesting takeaway? Which topics are particularly interesting for the clients? Depending on the experiences on site and possible upcoming questions, further information is of course given and small topic-specific "workshops" or, if desired, low-threshold short consultations take place.

- **Counselling within the framework of outreach work**
  There are milieus in which prevention counselling has to take place almost exclusively on site, either because the target group hardly ever arrives at the advertised facilities (e.g. hustlers, but also sex workers from street prostitution areas and clubs) or because there is simply no facility, especially one with medical services, to which they could be referred. This is especially the case in rural regions and smaller towns and poses a particular challenge for the work, because medical care and social support are even more time-consuming to organise than with a counselling centre in the background. Under such conditions, therefore, some prevention messages and information must take place directly in the milieu. People who lead a double life or who have already experienced discrimination and fear new ones can also be reached better on site than in an institution.

Based on this consideration, complete on-site services have been established again and again and increasingly in recent years, i.e. even medical examinations or the taking of samples for laboratory diagnostics are offered on a mobile basis on site. Some positive examples show that this is possible.[5] The use of resources is particularly high here and the scope of medical examinations is limited depending on local conditions. It should also be noted that in many settings of the sex work scenes, individual confidentiality is difficult to establish and the right to self-determination with regard to the use of diagnostics cannot always be guaranteed either. On-site counselling options are usually limited in terms of subject matter and time, and there is usually no opportunity to have a trusting, longer counselling session in the sex work scene. A distinction must therefore be made between counselling services that can take place directly in the scene and those that should rather take place in the facility for professional reasons and to protect confidentiality.

[5] In particular, the AIDS support services are developing testing services that are carried out locally.

4 The Impact of Outreach Work on the Uptake of Services in the Health Department

Unfortunately, we do not have a systematic scientific study on the effects of outreach work on the use of services in facilities. However, experience reports from facilities of the ÖGD and specialist counselling centres for sex workers and our own smaller studies suggest that the use of social and health services is significantly influenced by outreach work. In the following we present an example:

As part of the investigation into the expansion of the restricted area of the city of Cologne (Steffan & Arsova Netzelmann, 2012), interviews were conducted with sex workers in the south of Cologne and social counselling in the mother tongue was offered to sex workers from Bulgaria, who formed the main group in the region affected by the expansion of the restricted area. Due to scarce resources, the social workers of the STI counselling centre of the city of Cologne can only be on site in the south of Cologne once a month. In the run-up to the survey of the sex workers, the activities were intensified, and more extensive and frequent street work assignments took place in the streetwalking areas of the south of Cologne. In addition, the team carrying out counselling and outreach work was expanded to include a Bulgarian-speaking social worker from another agency. These measures had a great impact: On the one hand, the level of awareness of the health office's services in the scene increased, and on the other hand, more women from this group visited the health office because of the possibility of counselling in their mother tongue. In particular, the number of Bulgarian sex workers who visited the STI consultation hours increased significantly as a result of the intensive outreach work; compared to the reporting year 2009, the number almost doubled. The need for (mother-tongue) social counselling and support also became apparent. In the newly established social counselling service of the Bulgarian social worker mentioned above, 87 sex workers made use of a total of 141 counselling sessions, accompaniment and support in the short model project period of less than three months. Overall, those seeking advice had little to poor knowledge of German and had little knowledge of the German health and social system. The importance of access to social counselling at the health office via outreach work is also shown by the reasons for counselling identified in the study.

Since the social counselling took place within the framework of the medical consultation hours of the STD counselling centre at the Cologne Health Office, health issues were at the top of the frequency list, followed by a lack of health insurance, precarious working conditions in sex work, the desire to leave prostitution, dealing with experiences of violence and abortions, as well as financial hardship and existential issues. The work of the social worker consisted of crisis intervention, psychosocial support, assistance in recovering personal documents, accompaniment to offices, etc. In many cases, interpreting was also provided. In many cases, interpreting was also provided, especially in the context of pregnancy conflict counselling, accompaniment to abortions, medical care and telephone contacts with doctors and authorities. In addition, the Bulgarian social counsellor also took on the function of linguistic and cultural mediation both in the medical STI consultation itself and in the referral to other specialist departments of the health department, e.g. in pregnancy conflict counselling.

Since in almost all sex work scenes in Germany the proportion of migrants is large, the use of social workers with language skills in the languages of the most common nationalities is very important. The countries from which migrants come to Germany and become active in sex work here depend on global and European-wide migration movements. Over the years, many "waves" of migrants from different countries have been observed in sex work (MEGEPA, 2014). It will therefore not always be possible to keep social workers with appropriate language skills available for this task in the short or medium term. In these cases, the use of specially trained cultural mediators is a good alternative; they can be employed as additional freelancers and can be deployed flexibly.

5 Conclusion

Within this framework, various topics could only be touched upon; an intensive reappraisal of the role of outreach work within the framework of health services could certainly contribute to fertilizing the discourse on an adequate structure of services. The relevance of this work cannot be stressed enough, especially in the context of the controversial debate on new legislation for the field of prostitution in 2015. Systematic outreach work in connection with offers in the ÖGD and in counselling centres for sex workers represents an important bridge between prostitution milieus and society and enables acceptance, integration and, if desired, a change with professional and personal reorientation.

For the sake of completeness, it should be mentioned here at the end that many important areas of work of social work in STI counselling in the ÖGD are missing in this article. In addition to test counselling and prevention, the areas of documentation and networking are missing. Internal cooperation between medical services and social work is also missing. However, the participatory design of this cooperation is crucial for the success of synergetic effects from both fields of work. Also the networking, which is intensively carried out by social workers as well as by physicians in this field of work in the ÖGD, is not the subject of the above descriptions.

But it is precisely in the diverse networks at the professional, municipal and state levels, as well as at conferences and congresses, that the expertise of the ÖGD in this field of work is used to publicise the needs of clients, to criticise grievances and undesirable developments, and to constantly adapt services and develop them further with the help of political decisions.

References

Bundeszentrale für gesundheitliche Aufklärung (BZgA). (2015). PQD Case Studies – Quality action: Improving HIV prevention in Europe. http://qualityaction.eu/wordpress/wp-content/uploads/2014/02/PQD-Case-Studies-FINAL.pdf . Zugegriffen: 29. Jan. 2015.

Gusy, B., Krauß, G. M., Schrott, G., & Heckmann, W. (1994). *Aufsuchende Sozialarbeit in der Aids-Prävention – Das Streetworker-Modell* (Schriftenreihe des Bundesministeriums für Gesundheit Bd. 21). Baden-Baden.

Howe, C. (2011). *Nachbarschaften und Straßen-Prostitution – Konfliktlinien und Lösungsansätze im Raum rund um die Kurfürstenstraße in Berlin. Im Auftrag des Bezirksbürgermeisters von Tempelhof-Schöneberg, unter Mitarbeit von Milena Sunnus*. Berlin.

Krisch, R., & Stoik, C. (2011). *Soziale Arbeit im öffentlichen Raum – Ein Glossar zur ressortübergreifenden Rollen- und Begriffsklärung in Wien.* Verfügbar unter http://www.sozialraum.de/soziale-arbeit-im-oeffentlichen-raum-glossar.php . Zugegriffen: 8. Feb. 2015.

Leopold, B., Elfriede, S., & Nikola, P. (1994). In *Dokumentation zur rechtlichen und sozialen Situation von Prostituierten in der Bundesrepublik Deutschland* (Hrsg. Bundesministerium für Familie, Senioren, Frauen und Jugend. Schriftenreihe Bd. 143). Kohlhammer.

Ministerium für Gesundheit, Emanzipation, Pflege und Alter des Landes Nordrhein-Westfalen (MGEPA). (2014). Der Runde Tisch Prostitution Nordrhein-Westfalen – Abschlussbericht. Bd. 148. http://www.mgepa.nrw.de/mediapool/pdf/emanzipation/frauen/RTP_Abschlussbericht.pdf . Zugegriffen: 8. Feb. 2015.

Robert-Koch-Institut (RKI). (Hrsg.). (2012). Befragung zum STI/HIV-Angebot und zur Datenerhebung in Gesundheitsämtern, 2012. http://www.rki.de/DE/Content/InfAZ/S/STI/Befragung_Abschlussbericht.pdf?__blob=publicationFile . Zugegriffen: 8. Feb. 2015.

Rogers, C. R. (1951/1991). *Die klientenzentrierte Gesprächspsychotherapie*. Frankfurt a. M.: Fischer. (Orig. ersch.: 1951 Client–centered therapy. Its current practice, implications, and theory. Houghton Mifflin.).

Schmid, P., & Keil, W. (2001). Zur Geschichte und Entwicklung des Personzentrierten Ansatzes. In P. Frenzel, W. Keil, P. Schmid, & N. Stölzl (Hrsg.), *Klienten-/Personzentrierte*

Psychotherapie. Kontexte, Konzepte, Konkretisierungen (Bibliothek Psychotherapie, Bd. 8, S. 15–32). Facultas.

Stiftung SPI (Hrsg.). (2009). Infoblatt Nr. 1 Schweigepflicht Anzeigepflicht Zeugnisverweigerungsrecht. www.stiftung-spi.de/download/sozraum/infoblatt_01.pdf. Zugegriffen: 31. Jan. 2015.

Steffan, E., & Arsova Netzelmann, T. (2012). *Erstellung einer Konzeption für psychosoziale Unterstützung und gesundheitliche Betreuung von Sexarbeiter/innen in konflikthaften Straßenstrichbereichen, Assessment der Situation der Straßenprostitution im Kölner Süden nach Einführung der Sperrgebietserweiterung am 1.5.2011*. Abschlussbericht im Auftrag und gefördert von der Stadt Köln.

Steffan, E., Rademacher, M., & Kraus, M. F. (2003). *Gesundheitsämter im Wandel*. Forschungsbericht 296, Sozialforschung. Bundesministerium für Gesundheit und Soziale Sicherung (BMGS). Berlin.

World Health Organization. (Hrsg.). (2002). SEX-RAR guide: the rapid assessment and response guide on psychoactive substance use and sexual risk behaviour. http://www.who.int/mental_health/media/en/686.pdf?ua=1. Zugegriffen: 16. Feb. 2015.

Elfriede Steffan holds a degree in sociology and heads the SPI Forschungszentrum gGmbH in Berlin. She conducts research in various fields on the topic of prostitution.

Tzvetina Arsova Netzelmann is a psychologist (M. A.) and works as a research assistant at the SPI Forschungszentrum gGmbH in Berlin. Her focus is, among other things, on the coordination of EU projects.

Sex Work and Social Work: An Instruction Manual

Jacqueline Suter and Melanie Muñoz

1 Sex Work Is Work

The importance of sex work for our society should not be underestimated. On the one hand, the countless sex workers and their equally countless clients testify to this, but on the other hand, the fact that even at the beginning of the twenty-first century, sex for sale still provides material for emotionally charged socio-political debates. For the understanding of this essay it is important to know that the discourse on sex work has not always been the same and that sex work as a social and individual reality is also incessantly changing. The term "sex work" is used deliberately because it refers to women, men, transvestites, and transgender people who – to varying degrees of self-directed or other-directed – engage in a trade, generate profits, earn money for themselves and their families, pay taxes, etc. What all sex workers have in common is that they perform their work with services aimed at the sexual stimulation or satisfaction of clients. Another point is that when society talks about "prostitution", "it usually does not mean a professional activity, but an affair that encompasses one's whole life: one does not work as a prostitute, one is a prostitute" (Hürlimann, 2004, p. 2). Society finds it difficult to accept that sexuality can be bought and that sexual services for payment are described as a job.

J. Suter (✉)
Bern, Switzerland
e-mail: jsuter@xeniabern.ch

M. Muñoz
Olten, Switzerland
e-mail: info@lysistrada.ch

M. Albert, J. Wege (eds.), *Social Work and Prostitution*,
https://doi.org/10.1007/978-3-658-37761-8_7

In Switzerland, sex work has been recognized as a reality on a legal-political-societal level for more than half a century. Despite this recognition, sex workers have few rights and are exposed to structural and social stigmatisation. Conflicting social signals are sent to sex workers: On the one hand, they are supposed to be responsible taxpayers and citizens; on the other hand, as providers of sexual services for remuneration, they are still labelled as second-class people.

Sex workers live in very different realities and bring with them very different prerequisites. It is by no means possible to speak of a homogeneous group, i.e. of sex workers. Some of them have freely chosen this work and have their own professional identity. Others pursue sex work because of precarious economic conditions. In some cases, third parties are involved who exploit the sex workers economically, sometimes by force. In most cases it is difficult to prove the existence of forced conditions, partly because in the self-perception of those concerned this exploitation is often not perceived as such. Especially the media and the abolitionist movement make it too easy for themselves when they stylize sex workers, especially women, exclusively as victims. Whatever the circumstances, motivations and motives of the sex workers are: sex work is a job in the middle of society. It is defined by its contradictory otherness and it is precisely by virtue of this otherness that it contributes to maintaining the "normality" of the prevailing social structures. This otherness (Augello, 2010, p. 249; Bastian & Billerbeck, 2010, p. 45 ff.; Löw & Ruhne, 2011, p. 149 f.) affects not only the image of sex work in the public eye and among clients, but also the self-perception of sex workers. The social environment in which sex work takes place is highly complex. If social work wants to act professionally and successfully in this field, and if it wants to support sex workers as advocates, then it must unconditionally recognise and respect sex work as work.

2 Three Ways of Looking at a Phenomenon

The political attitude of the countries of this world to the phenomenon of sex work fluctuates between three systems. In countries with prohibition, sex work is legally forbidden. A system of regimentation regulates sex work; it is hoped that this will lead to better control and thus simplified intervention, be it repression or prevention, to the benefit or detriment of sex workers. Abolitionism has the declared aim of abolishing sex work, because it sees women as victims of society and is thus against both the repression of sex workers and the regulation of sex work. The modern variant of abolitionism is the so-called Swedish model, which bans the purchase of sex and thus punishes the johns, while leaving sex workers alone under criminal law. Switzerland is currently moving differently depending on the canton and accentuated between the poles of regimentation and abolitionism, with the latter gaining importance mainly as a public discourse in the media and politics. The regulatory approach is currently still clearly the mainstream, which means that sex work continues to be accepted as a social reality in Switzerland. From a social work perspective, this is to be welcomed in principle, although a warning must be issued against pushing sex workers even further into the margins of society with increasing control and registration measures, all the more so as these measures are often due to migration policy considerations.

3 Legal Framework in Switzerland

The legal framework of sex work in Switzerland has been comprehensively presented by Brigitte Hürlimann in her standard work "Prostitution - its regulation in Swiss law and the question of immorality" (2004). Sex work has been permitted in Switzerland since 1942 and is considered a form of economic activity. Those wishing to engage in sex work must be 18 years of age and must comply with the regulations in force in the cantons and cities regulating the practice of sex work. Persons from abroad must also comply with the provisions of the law on residence and work. This means that they need a settlement permit or a residence permit or a work permit. This is made dependent on various requirements or conditions and in some places is simply refused for sex workers. Despite legalisation, contracts between sex workers and establishment operators are still considered immoral. To make matters worse, due to this immorality, client wages owed by sex workers cannot be claimed to this day. However, political initiatives are currently pending on this issue. As already mentioned, the regulation of the sex trade falls under the competence of the cantons and/or cities. In federalist Switzerland, this therefore means a real diversity of different laws, ordinances and decrees. This is worth mentioning because sex workers are highly mobile, work in different places and it is a great, almost unmanageable challenge for them as well as for social workers and counsellors to gain a legal overview of this Swiss diversity. Because what applies and is the rule in Bern, for example, can look completely different in Zurich. Practicing sex work legally means not only having the appropriate residence and work papers and complying with the various regulations on prostitution (such as the controversial registration requirement for sex workers), but also following cantonal and municipal guidelines. These include, for example, trade police, building and neighbourhood regulations. If you as a social worker want to give good and correct advice to a sex worker who is striving for legality, you have to be familiar with the applicable laws and show perseverance during this administrative hurdle race. In Switzerland, one finds oneself (and this applies in particular to migrants) unpleasantly quickly in the illegal or at least irregular area of sex work.

4 Sex Work and Migration

Female sex work is difficult to consider separately from the phenomenon of migration. The fact that this was already the case in the past was noted by Anita Ulrich in her work "Brothels, Streetwalkers and Bourgeois Morality in the Belle Epoque", published in 1985 and based on data collected in 1870 and 1890: "The majority of brothel girls came from abroad, recruited mainly from the rural areas of southern Germany (Bavaria, Baden, Württemberg) and Alsace. The second largest contingent came from - likewise rural - places in the canton of Bern, Aargau, from the interior and from eastern Switzerland. There was only a very small proportion of Zurich girls and girls whose place of origin was very far from Zurich. In addition, there was a strongly fluctuating group of prostitutes from non-German-speaking areas, for example Hungary, France, Italy, exceptionally even Algeria" (Ulrich, 1985, p. 15). The sex workers arrested on the streets of the city of Zurich in 1881 and 1882 were also almost exclusively migrants from the countryside (Ulrich, 1985, p. 38). It is obvious to assume that this fact has structural reasons whose logic reaches far beyond individual fates. Female sex work cannot and could not be described separately from the phenomenon of migration of rural lower-class women into the urban service sector.

In contrast to the nineteenth century and most of the twentieth century, when female labour migration in Switzerland was very regional, the vast majority of sex workers in Switzerland today come from more distant parts of the world. Typical countries or areas of origin are, for example, the Dominican Republic, Brazil, African countries and Thailand. After 1990, the countries of the European Union (2004 Hungary, 2007 Romania and Bulgaria) have joined the ranks. As a result, the legal situation of women working in the sex trade has become unevenly more complex, as the legal conditions under which women can engage in sex work also vary widely depending on their residence status. The legal framework for foreign nationals living in Switzerland is governed by the Foreign Nationals Act, which was approved by the Swiss electorate in a referendum in 2006 and came into force on 1 January 2008. Basically, a distinction is made between persons from the countries of the European Economic Area (EU and EFTA) and persons from so-called third countries. In principle, persons with a permanent residence permit are treated the same as persons with a Swiss passport in terms of employment law.

Since the Agreement on the Free Movement of Persons between Switzerland and the European Union came into force in 2002, EU citizens have gradually been granted the right to work in Switzerland. However, this freedom of movement is undermined in the field of sex work because the migration authorities of the cantons and, in some cases, of the municipalities and cities have the authority to impose their own conditions on the residence and working conditions of sex workers. This repeatedly leads to absurd and contradictory situations. For example, sex workers from the EU or EFTA must either overcome high bureaucratic hurdles as self-employed workers or present an employment contract with an establishment operator as self-employed workers – the latter, although in Switzerland the "promotion of prostitution" should actually be prosecuted by law.

People from third countries have no possibility to legally engage in sex work in Switzerland, except by marrying a Swiss national or a person established in Switzerland and/or EU nationals living and working in Switzerland. The so-called cabaret dancers are an exception. Here there is the so-called "residence status L", which is linked to monthly work contracts. During this time, the dancer does not have the possibility to pursue other work. The maximum duration of her stay is eight months in one year. Incidentally, the study "Working and living conditions of cabaret dancers in Switzerland" by Janine Dahinden and Fabienne Stants (Dahinden & Stants, 2006), published in 2006, provides well-founded information on this subject. Within this framework, only a small glimpse of the complex situation in which migrant sex workers find themselves is possible. What seems essential here is to point out once again the particularly vulnerable situation of third-country women who, as sex workers, either live and work in illegality as "Sans- Papiers" and are bound to monthly contracts as dancers or local residents willing to marry. All these cases require very special attention from sex work counsellors.

5 Social Work in the Field of Tension of Sex Work

A Case Study

In order to present the everyday life of social work with sex workers from the perspective of practice, a case study will be presented here at the beginning. In the following, this case serves as a kind of instruction manual for social work in the sex trade. It discusses methods and areas of work, the demands on social workers, and the financial and institutional context of the work. The main protagonist of this example should be called G. here. She has a Thai and a Swiss passport. Thanks to the method of outreach work, regular contact has existed for some time. G. has been living and working in Switzerland for 25 years. First she had run a salon in a small town, which had burnt down, and most recently she ran a salon in B. with a changing number of working women. In this establishment all the sex workers had the same nationality. There were salons with different female bosses. G. was one such boss. The social workers always had consultations on site, be it with the women working there or G. as a boss had questions. The counselling centre in B. had never come to G.'s attention.

The social workers knew that there were women working for G. who did not have legal residence status. The aim was to get in touch with the women working there, to give them the information they needed to get out of their dependency, which stemmed from their illegalisation. The social workers had intensive conversations with G. about the situation of the sex workers and confronted her again and again with the situation of the illegalised women working there. G. knew very clearly that if a woman had wanted to testify against her, the counselling centre would have supported her. G. meanwhile had great confidence in the social workers and their work and allowed them to speak alone with the sex workers present. She informed the social workers each time new women were brought into the house and provided access to the other parlours. The team of social workers always travelled in pairs; one spoke to G., the other was able to talk to the women in their mother tongue in the next room. After some time, a change occurred: one day the house no longer conformed to zoning requirements, namely, it was used mostly for commercial purposes instead of as living space (for example, in four-room apartments, three rooms, sometimes all rooms, were used as workrooms; sleeping was done in the kitchen on a rotating basis, since it was a twenty-four-hour operation). After a lengthy legal dispute between the landlord and the city, the Federal Court ruled that the entire house could only be used for living and set a deadline for returning it to this use.

Thanks to good contacts with the sex workers, a meeting of the "bosses" involved could be organised. Almost all of them were over 55 years old and had never done any work other than sex work for more than 20–30 years, had poor German language skills and were partly illiterate. In total, ten women were involved. Options were discussed together. It was decided to hold a plenary discussion with all ten women in addition to the individual discussions in order to make joint demands on the landlord. As the women all had their centre of life in this house, there had to be a possibility that the tenancy agreements could be changed to the extent that a rent that could then be taken over by social services would exist, at least on paper, even if the landlord might demand black money for it. However, it was clear to all that the women would not be able to find work or any other accommodation any time soon, if ever, as they had no income. Because the repossession of the house deprived them of their livelihood. So the only solution was for them to register with the social services. After negotiations with the landlord, the rental contracts were changed in the presence of the social workers. As a final show of force, the landlord agreed to hand over the contracts to the tenants, but not to sign them until exactly 11:55 p.m., the legal deadline for signing. So the social workers spent the evening in the "whorehouse" until the last minute. An appointment was made with all ten women at the social services office. The suggestion to register them all at the same time was at first met with resistance by the social services, but they were eventually persuaded by practical arguments. The social workers put together all the documents for each former sex worker individually, so that things would go faster at the social services. Each of them received support: for some an early retirement could be envisaged, others went back to their country of origin and still others remained with the social services.

Since the housing situation was only considered an emergency solution, the counselling centre staff looked for another apartment with G. as soon as possible. Since G. was illiterate, she had to be accompanied regularly to view apartments. Her only social network in Switzerland consisted of her community. She needed support to be able to read housing advertisements, to write applications and forms, etc. She also needed help in finding a new home. Thanks to the cost approval of the social services, she found an apartment quite quickly. Everything was set up and the counselling team was able to withdraw after intensive support. It was agreed with G. that she would not sign anything she did not understand, and that in this case she would contact the counselling centre in advance by telephone or send a copy of the respective letters. After a year she got in touch: She had married in the meantime, but was now accused of bigamy. She had married a Turk in Turkey and only had Turkish marriage certificates. This man had been sentenced to 16 years in prison for various offences. Her new husband, also Turkish, was arrested and had to serve his sentence in the same prison where her first husband was serving time. There the two men found out that they were married to the same woman, whereupon husband number one filed a complaint. G. then explained that since she had never registered as married in Switzerland, this marriage was not valid. She had married in Turkey and not in Switzerland. These conversations were very emotional and loud (G. is a resolute personality): The first man was her great love and she had only married the second one so that she would get money and not be alone in her old age. What was to be done? Firstly, a lawyer had to be organised to initiate the divorce and to file an application for free litigation. Secondly, G. had to be informed about the consequences of filing a complaint, namely that the migration authorities would consider her second marriage as a sham marriage, since bigamy is forbidden in Switzerland.

The divorce from the first husband could be carried out; with tears, as it was her great love, but also with humour – for G. was well aware that she loved a "bandit" to whom she had given all her money and jewellery for years, whom she had hidden from the police and who had pressured her to take in illegalised women in her salon. She had also been married to him twice in Turkey, the first time under his brother's name. She had divorced him and married him again, under his real name. He still had power over her even from prison, where she visited him regularly. G. now wanted to divorce her second husband as well. After deliberations on her part, she had found out that he was of no help to her, since he was also in prison and because she had had to endure a police search of her house because of him, about which she had been very ashamed. This divorce was also initiated. Afterwards, long and intensive counselling sessions were held again with the aim and the request that she should not get married, that the counselling centre would no longer be prepared to become active on the subject of her marriage.

The Sex worker's Perspective

This is roughly how the reality of a social worker might present itself from the perspective of a sex worker.

Why should I trust a social worker? She gets paid every month. Five weeks a year she is not to be seen, she spends her holidays with her family, her life partners and friends, to whom she often and gladly tells about her work in the sex trade. I, on the other hand, as a sex worker, sit on my hands, I don't want my family and my children to know anything, or at least no details. No one cares about my retirement plan, I have to take care of it myself, better yet, not think about retirement at all. Why should I trust a social worker? She is on a first-name basis with the police and other authorities, with pimps, brothel operators and, last but not least, with my competitors. She is too young or too old, has no idea about "real" life, what does she want to tell me? Isn't it the case that I'm still waiting for customers on the street or in an establishment when she's already enjoying the end of the day or the sleep of the innocent?

Gaining the trust of a sex worker is not easy. Often, before a real relationship can develop, there is a long preliminary phase in which the sex workers test the social workers for their trustworthiness and want to find out whether they are taken seriously as individuals, as women and as sex workers. They want to provoke, for example, through emphatically sexualised verbal or non-verbal behaviour, through racist, xenophobic slogans or the comment that counselling services are only for foreigners anyway.

The aim of any relationship building in social work in the sex trade must be to pick up the sex workers in their lifeworld, to build bridges, to make them understand that social workers basically do not want anything from them, that direct contact with them does not obey any economic rules and can be particularly valuable for this very reason, a mirror, so to speak, in which sex workers can look at themselves from the outside and gain insights into other realities. Of course, this is only possible if social workers and sex workers find a common language, in which the technical terms of the social workers have only limited space if social workers unconditionally accept the life worlds and decisions of sex workers, which may seem strange to them, without fading out emotions (neither those of the sex workers nor their own). In order to appear authentic and thus credible towards sex workers, social workers must also be able to directly and bluntly bring up what seems strange to them. They should talk about power and dependencies without giving up the ability to laugh together with the sex workers about the absurdities of life.

Qualifications of a Counsellor/Social Worker

Social workers in sex work should – which unfortunately is by no means self-evident in certain, especially religiously motivated offers – have no problem with sex work as a social phenomenon. For this work, a certain degree of acceptance, respect and interest as well as curiosity for people who do not conform to the "norm", who – as is common among sex workers – have a verbally and relatively rather offensive approach to sexuality, or whose gender is ambiguous or changing, as is the case with transvestites or transgender people, is required. Profound knowledge of legal and health issues are fundamental requirements for all social workers who work with sex workers. In everyday work with them, empathy, pragmatism and the ability to deal with conflict are important prerequisites. In order to pursue their heterogeneous interests for an improvement of their working and living conditions on a strategic level with the authorities, the operators, the police, etc., qualities such as strategic thinking, negotiating skills, persuasion and perseverance are indispensable. Again and again, the question must be asked who are the allies with whom a goal can be achieved. It is often necessary to act with great determination and to keep one's own limits in mind. Last but not least, it is necessary to endure the imperfect world of the sex workers, who usually only contact the representatives of social work when the fire is already in the roof.

Without rigorous self-reflection (e.g. in the context of supervision and/or intervision) and resource-oriented thinking and acting, it is not possible to do professional work in the milieu of sex workers. It requires the willingness and awareness to work in a legal and ethical grey area, where there is no black or white, no clearly separated perpetrator and victim roles, where every case and every situation is different. Sex workers need to be extremely resilient (even in cases that border on the intolerable), persistent, empathetic and patient, and able to tolerate the fact that it can sometimes take ten years for sex workers to tell their "true" and "whole" story. Other requirements are extensive communication skills, language skills and the ability to think and act in a complex and networked way. It requires the willingness to go the extra mile, to pull all-nighters again and again, to be absolutely reliable without forgetting to demand obligations from the other actors in the field of work as well. All this is not possible without a high level of social competence, not only towards the sex workers, operators, etc., but also towards the colleagues at work. For reasons of psychological hygiene alone, long-term, healthy work in this field is unthinkable without complicity among social workers, without mutual gossip.

6 Methods

The most important complementary methods are outreach work, inpatient counselling and online counselling in order to reach sex workers as comprehensively as possible. With regard to the duration of contacts, short and long-term counselling as well as crisis intervention can be mentioned. Due to the aforementioned mobility in the sex trade, short-term counselling and crisis intervention are the most common. The approach is always life-world and solution-oriented. The sex workers themselves are experts in their working and living environment and, in the best case scenario for social work, can be persuaded to recognise and use their own resources. Continuous counselling at the sex workers' place of work can also be of decisive importance, as the case study G. has shown. A special case in social work in the sex field is the use of mediators. The idea of this model is to maintain contact with sex workers from other language regions with the help of women who have the same mother tongue. Women who work in the sex trade need information that is not based on hearsay. The sometimes lacking professional distance of the mediators used undoubtedly also has its disadvantages, but these are always outweighed by the unsurpassed low-threshold nature. Working with mediators requires professional and intensive support. This extremely low-threshold model was launched in Switzerland 20 years ago as a pilot project and is now being implemented nationally by various organisations.

Areas of Work in the Counselling of Sex Workers

Efficient services in the area of the sex trade can only be provided by social counsellors if they think in a networked way and work in a networked way. Networked thinking is much more the case in this area of work than in other areas of social work, because the target group largely has neither a Swiss passport nor a permanent residence permit, is extremely mobile and very strongly connected to their regions of origin. Thus, one of the most urgent networking tasks in social work in the sex trade is contact with organizations in their countries of origin, in neighbouring countries, and with members of the corresponding communities in Switzerland. The mobility that characterizes the sex trade also affects the places of work in Switzerland. Here, partly because of the different legal provisions in the cantons, it is essential to maintain contacts with other organisations at national level and to be in contact with politicians, authorities and decision-makers at communal, cantonal and national level, which can only be achieved throughout Switzerland by networking the various organisations, most of which have very local roots. The Swiss network ProCoRe (prostitution collective reflection), which has been working since 2000 for better working and living conditions for sex workers in Switzerland, is nationally organised and oriented. This association of professional organisations guarantees their indispensable professional exchange throughout Switzerland. Two main areas of work have been established in practice: Legal and financial counselling on the one hand, and health issues and psychosocial counselling on the other. Legal advice is sought on topics such as immigration law, family law, tenancy law, employment law, social security law and victim assistance law, while financial advice focuses on tax returns, debt restructuring, budget advice and business plans. Health issues include women's health issues, sex reassignment surgery, STD prevention, and addiction issues. Psychosocial counselling can concern sexual violence and other forms of violence, but also family conflicts in Switzerland or in the home country, "homesickness" and other psychological or social problems. The fields of activity of social counsellors in the sex field are wide open on all sides. There is always and everywhere room for manoeuvre. The maxim is: at best, there is always a way to represent the sex worker's concerns and act accordingly. In public and political work, it is also important to always be able to react innovatively and proactively to new or difficult situations.

The Other Side of the Sex Trade: Clients of Sex Workers, Operators, Etc.
Social workers in the sex trade operate in a social field in which various actors are on the move. Advocacy for the interests of sex workers is at the heart of their work. This does not mean that professional contacts with other actors are superfluous. Contact with johns and operators is particularly important when it comes to understanding how the world of sex work works and being able to get in touch with as many sex workers as possible. It is not uncommon for johns, friends of sex workers or other actors (including possible "pimps") to contact counselling centres and institutions to seek information on behalf of women or to denounce abuses and misconduct by police or authorities. This results in a high importance of john contacts for social work in the sex trade. However, it must also be taken into

account that supportive actions by clients are in many cases well-intentioned, but in the end are often based on their own self-interest, which in turn can lead to (new) dependencies for sex workers.

Despite these limitations, it should be noted that clients are basically the actors who, apart from the sex workers themselves, are most aware of their working conditions and can therefore, with the appropriate awareness and interests, also recognise grievances and problems, address women about them or offer them help. It is part of the professional attitude to deal with the men in the sex work milieu in a non-judgemental way and to lower their inhibitions to get in contact with the specialist and counselling centres that are considered to be male-hostile feminist strongholds. Similarly significant is the contact with the operators of establishments as well as with men and women who exercise control and boss functions in such places or on the street prostitution. As social workers, it is also important to open up professional access to them, to build and maintain relationships, whereby one of the most important problems in this context is the power imbalance, at the lowest end of which are often the sex workers themselves. As a result, they are not allowed to talk to social workers without permission from their superiors. Examples of this are clubs and brothels, which one could not even enter without the permission of the operators or their bosses present on site.

Threefold Mandate: Institutions – Sponsors – Financing

In Switzerland, with its small-scale political organization, social work in the social sector is inevitably also organized on a small scale. Today's services can be divided into different groups in terms of their history and legal form. The first contact and counselling centres, which were explicitly and exclusively aimed at sex workers, were set up in the 1980s on the initiative of politically committed street workers, and at the beginning often had so-called drug prostitution in particular in mind. These centres were all organised as independent associations. These include Aspasie in Geneva, in which sex workers have played a leading role since its foundation in 1982, which unfortunately remains an exception in Switzerland to this day, the Xenia specialist and counselling centre in Berne, founded in 1984, and other associations founded in the 1990s, such as Lysistrada in Olten (canton of Solothurn) since the beginning of the 1990s or Fleur de Pavé in Lausanne (canton of Vaud) since 1996. Another group of counselling centres has emerged from pre-existing church or non-profit services, such as the counselling and contact centre Isla Victoria, which has been active in Zurich and Winterthur since the 1990s and emerged from the Midnight Mission of the Evangelical Reformed Church in Zurich and is now under the auspices of the Zurich City Mission. The Aliena counselling centre, established in Basel in 2001, has a similarly long history and is run by the non-profit association Compagna, the former "Friends of Young Girls", which was already active at the end of the nineteenth century. As a form of organisation, direct organisational affiliation with the state should be mentioned, as is the case with the Flora Dora service in the city of Zurich, which is present on the Zurich Strichplatz, or the Maria Magdalena service in St. Gallen. The services and counselling centres mentioned so far are united by their social work and advocacy approach to sex workers, their membership of the umbrella organisation ProKoRe and the fact that they are either directly dependent on the state, as in the case of Flora Dora and Maria Magdalena, or indirectly dependent on state agencies through service agreements with cantons or cities in the case of the other centres. Activities from the environment of the free church, evangelical offers will not be discussed here, as these mostly do not carry out professional social or pastoral work, but rather missionary work, and do not perceive the sex workers as acting subjects, but rather as victims. The current state of affairs is that most social services in the field of sex work receive state funding, which confronts the professional and counselling centres with various problems:

A first problem area is the administrative demands on modern, public organisations, which are also increasingly demanded of services in the social sector that fulfil public service mandates. Here, in recent years, under the auspices of the New Public Administration, there has been a massive increase in work, especially in the area of statistics management and time recording systems. The measurability of results, especially in social work in the sex sector, is sometimes very difficult, if not actually impossible. On the one hand, it sometimes takes years and countless re-contacts before a sex worker uses a service at all. On the other hand, there are consultations that last for years without visible results. Another problem is the pressure for certification, which can plunge small associations in particular into financial abysses, as certification is not only very time-consuming but also costly. All in all, it should be noted here that small organisations in particular, which often live from the voluntary, unpaid work of board members and from no more than one hundred to two hundred job percentages, are always at the limit of their capacity.

Another problem in this context is the fact that sex work has just now become an important political issue again. The newly sparked prohibition debate or the Swedish model of punishing johns, the discussions about visible sex work, for example in the form of street prostitution, and last but not least the permanent scandalisation of sex work through media coverage mean that social services in the sex work sector are on the one hand exposed to frequent observation and criticism in the public sphere, but on the other hand are also subject to strong demands from the media and politics. All of this work – let us be blunt here – is usually extremely frustrating in the context of political commissions and legislative consultations.

While everyone talks about the protection of women, in reality this protection is subordinated to migration policy and the need for security and order, and it is at best minimally compensated or even explicitly prevented by service agreements. In the case of state-run services, it may seem appropriate that employees are not allowed to carry out political work during their working hours. However, small organisations, whose advantage lies precisely in their independence (which should remain sacrosanct despite performance agreements), can at worst be silenced relatively quickly by threats to cut services. It is probably no coincidence that independent specialist and counselling centres are kept small financially by politicians, because this way their room for manoeuvre can be better kept under control.

A third problem relates to the tension between legality and illegality. The working methods in the field of sex work are rather unconventional and often situations have to be dealt with that take place in social grey areas. Professionally active social workers must always critically question themselves in this field of tension and comply with the mission and the declared attitude of the organisation, namely to act as advocates for sex workers. The tasks of the staff are becoming more and more complex, but not enough finances are made available. This raises the question of the quality of social work: service by the book or commitment? What form of social work in the sex trade does the state want to afford? And what form can and does one want to afford as a social worker?

7 Conclusion: What Does It Do to Me as a Social Worker in This Field?

Social workers in the sex industry are exposed to various influences and interests. Somewhere between the interests of sex workers, state service recipients or funders, police, politicians, administrators, media people, clients, brothel operators, tax authorities and residents, they have to find a professional way and are forced to constantly review and reflect on their own values and moral concepts.

Various questions come to the surface – questions that are always relevant day after day: How do I muster the energy to keep tilting at windmills? Do I still recognize my limits? How can I stand going into meetings with authorities knowing that clients of sex workers are sitting there, acting and speaking against sex work? What kind of male image does this work present to me and what impact does or could this have on my personal life? Does working in an extremely sexualised environment affect my own sexuality?

References

Augello, E. (2010). „Für Frauen bezahlt man sowieso.“ Prostitution – ‚Normalität’ und konträre Konstruktionen. In T. Benkel (Hrsg.), *Das Frankfurter Bahnhofsviertel. Devianz im öffentlichen Raum*. Verlag für Sozialwissenschaften.

Bastian, N., & Billerbeck, K. (2010). *Prostitution als notwendiges Übel? Analyse einer Dienstleistung im Spannungsfeld von Stigmatisierung und Selbstermächtigung*. Tectum.

Dahinden, J., & Stants, F. (2006). *Arbeits- und Lebensbedingungen von Cabaret-Tänzerinnen in der Schweiz* (SFM-Studien 48). Neuchâ tel.

Hürlimann, B. (2004). *Prostitution – ihre Regelung im schweizerischen Recht und die Frage der Sittenwidrigkeit*. Schulthess.

Löw, M., & Ruhne, R. (2011). *Prostitution. Herstellungsweisen einer anderen Welt*. Suhrkamp.

Ulrich, A. (1985). Bordelle, Strassendirnen und bürgerliche Sittlichkeit in der Belle Epoque. Eine sozialgeschichtliche Studie der Prostitution am Beispiel der Stadt Zürich. *Mitteilungen der Antiquarischen Gesellschaft in Zürich* 52 (3). Zürich.

Jacqueline Suter is a counsellor and head of health and AIDS prevention at XENIA Fachstelle Sexarbeit, Bern, Switzerland. She is a board member of ProKoRe a nationwide network of organisations and individuals that represent the interests of sex workers in Switzerland and is particularly committed to improving their living and working conditions. She has been working and researching in the field of sex work for 18 years.

Melanie Muñoz is coordinator and advisor at the association “Lysistrada”, which has been working since 2000 for better working and living conditions for sex workers in the canton of Solothurn in Switzerland. She is a board member of ProKoRe, a nationwide network of organisations and individuals that represents the interests of sex workers in Switzerland and is particularly committed to improving their living and working conditions. She has been working in the field of sex work for 12 years.

Social Work with Prostitutes in Zurich: Illustrated by the Experiences of the Isla Victoria Counselling Centre

Regula Rother

1 Introduction

Prostitution is a topic which on the one hand is very well known and always and everywhere arouses high interest. On the other hand, it lies in the dark and it seems as if there is a world shrouded in mystery, far from the normal citizen and carried by male citizens, forbidden and never to be thought away. For the main actors – namely johns, prostitutes and pimps – remain in anonymity. Reactions and actions towards prostitution could not be more different: from total rejection to recognition as normal work, from pity for the poor women to condemnation of all male sexuality. When I first looked at the literature,[1] I noticed that there are always attempts to analyse the "oldest trade in the world", even to make it understandable, or at least to grasp the special problems of the trade and to show the reactions of society in different times. Very little is written about punters. They are the great unknowns in the business, although their number (one fifth of Swiss men) is known. The prostitutes themselves are usually reduced to their deficits or special problems,

[1] See, for example, Agustín, 2005; Bugnon et al., 2009; Dücker, 2005; Women MA57, 2013; Ulrich, 1985.

R. Rother (✉)
Zurich, Switzerland
e-mail: regula.rother@stadtmission.ch

M. Albert, J. Wege (eds.), *Social Work and Prostitution*,
https://doi.org/10.1007/978-3-658-37761-8_8

such as health, experience of violence and mental disorders. To see them as normal people like you and me is rare.

Considering them as normal people is of course related to the anonymity sought by the prostitutes themselves. Authentic statements about how they really are and how they live are still difficult to obtain today. Interview-style surveys are rarely possible, and when they can be conducted, they must be listened to extremely critically. This is because it is one of the main competencies of sex workers – and indispensable for their profession – to sense the expectation of the counterpart and to precisely align themselves with it. For this reason, their answers are to be evaluated extremely cautiously in principle and even more so in scientifically oriented surveys. In short: We are dealing with a phenomenon that is very well-known and at the same time very obscure.

In its call for papers for a conference on "Paid Desire" planned for November 2014, the "Network for Sciences of the Psyche - Entresol" soberly and precisely draws attention to the fundamental issues behind the discussions. It asks fundamentally about the "mechanisms of negotiation" when different discourses and value systems clash. "Firstly, it is about the 'right' (female/male) sexuality, secondly about the problem of alienated labour, and thirdly about injustices in a globalised world" (Entresol, 2014).

In my experience, the meanwhile established social work in the red light milieu is indeed mainly confronted with the social exclusion tendencies, but no less with the social voyeurism and the criminalization of sex work. It, on the other hand, seeks to promote women's autonomy, dignity and skills. I see, as can also be read in recent literature (cf. Galuske, 2007, p. 50 f.), professional social work in principle and therefore also with sex workers in the following tension: On the one hand, it means building up a relationship of trust with the prostitutes on the inside. Only then do the real stories become visible and only then can effective solutions be tackled in the interests of the prostitutes. On the other hand, social work in prostitution requires committed political work and networking with other institutions on the outside.

In the following, I focus on three aspects: firstly, on the situation of sex workers as migrants; secondly, on the experience and handling of power and powerlessness and this with regard to both sex workers and social workers; and thirdly, on the modified application and reflection of methods specifically in the work with prostitutes. At the end I formulate consequences for the future.

My presentations are based on my five years of experience in a private counselling centre in Zurich, Isla Victoria. This is a sub-project of the Zurich City Mission (ZSM), which I have been running for five years. I am therefore sending ahead an account of the situation in Zurich with regard to prostitution, in order to make the challenge of our counselling position Isla Victoria easier to understand. Already in 2011, we also had a professional exploration prepared for our specialist Isla Victoria office, which I can refer to, among other things, in the following until the end of 2011 (Zürcher Stadtmission, 2011).

2 Current Situation in Zurich

The Zurich City Mission and the Isla Victoria Outreach Clinic
The Zurich City Mission (ZSM) looks back on over 150 years of diaconal work with women in the sex trade. The beginning was made by a courageous rich citizen who founded a home for fallen girls. Today we provide professional and pioneering social work in the city and canton of Zurich with a counselling centre, a meeting place and outreach work. We counsel around 3700 women per year and make 23,000 initial contacts. We always try to react immediately to the ever-changing situation with small projects. The initiatives usually come from the female staff, who have their finger on the pulse of what is happening. Our budget is very limited and many things that are necessary we cannot tackle for financial reasons. Our political work is now receiving a lot of attention from the media, often as a counter-voice to the state's political stance.

Political Environment
In 1897, prostitution was banned in Zurich for a short time. "Prostitutes were prosecuted by the police, imprisoned, put in asylums or expelled" (Ulrich, 1985, p. 135). But as early as 1904, the "anti-morality initiative" was rejected by a two-thirds majority and prostitution was legalized (Ulrich, 1985, p. 135). Meanwhile, throughout Switzerland, prostitution is protected by the Federal Constitution under the freedom of gainful employment. In Austria and Germany the legal situation is similar, while in Sweden, on the other hand, prostitution is prohibited (cf. Frauen MA57 2005). The street prostitution at Sihlquai in Zurich became more and more a public nuisance from 2008 onwards and the pressure on the city government to bring about a change was very high. The Zurich City Council decided to draft a Prostitution Trade Ordinance (PVGO). This was enacted by the city council on 7.3.2012 (Stadt Zürich, 2012). The ordinance regulates street prostitution and salon prostitution in the city of Zurich. Three street sections are designated for street prostitution. In order to close the street prostitution at Sihlquai, a so-called street prostitution area (boxes) for car prostitution was created as an alternative. This is operated and supervised by the city (Social Department) itself. In order to operate

a salon, a permit is required and various conditions (e.g. of a structural nature and specifications for operation) must be met. Women who wish to prostitute themselves on the street in Zurich must apply for a police permit in person. The control body and licensing authority is the Zurich City Police. The ordinance has been in force for two years and the street prostitution area for one year.

Impact of the PVGO on the Work of the Isla Victoria Outreach Clinic
Already at the beginning of the discussions about the designated street-prostitution area we brought in our concerns: Not all sex workers want to and can work at the location. Women who do not meet all the requirements (especially if they do not have health insurance) cannot prostitute there because they do not receive a permit and they are thus pushed into illegality. Our main criticism, however, was directed against the detailed ordinances that made it much more difficult for sex workers to work. For example, until the introduction of the ordinance, two to six women could easily operate a salon. With the new ordinance, this is no longer possible. Many women cannot make the huge administrative effort, nor can they afford a lawyer. There was a great deal of additional police enforcement used after the ordinance went into effect in the traditional red light district of Circle 4, where there are many contact bars, to stop the now illegal hooking on the street. As a result, women, afraid of being checked, at times stopped going to the counselling centres altogether. They were now fined for even approaching customers. In the meantime, the police checks have decreased again, but the difficult situation for the women has remained. In particular, they can hardly find rooms or flats for their work. We notice that now considerably more women visit our lunch table in Isla Victoria. Perhaps an indication that they are earning less? We also meet more women who work without a permit, and it is clear that they do not turn to a government agency.

The Social Discussion in Zurich
In the last few years, the prostitution on Sihlquai was increasingly used by women from Hungary. These women were very young and had a more aggressive way of attracting customers than the resident women. The Hungarian women hardly spoke German, they were often guarded by pimps and it turned out that many came from a very poor village – often a Roma village. A pimping trial in Zurich that dragged on for years particularly boiled up this issue and in the end some Roma pimps were convicted. A factual and professional discussion was soon no longer possible and the two camps, namely prohibition of prostitution here and sex work as normal work there, were quickly formed. The discussion in Zurich is very similar to that in other European cities: Feminists want to protect sex workers from bad men and punish the latter for their behaviour; middle-class women's associations find sex

work unworthy of a woman and want to ban prostitution; religious circles want to save women. Another group, on the other hand, declares sex work to be a legitimate job, like any other; professional bodies stand in advocacy behind sex workers and speak on their behalf for the legalization and equality of the trade with other professions. Time and again, sex work is placed in very close relation to human trafficking. Of course, there are such serious crimes as well, any one of which is too many. However, we cannot agree with the often expressed opinion that most sex workers are trafficked women. Our experience shows that most women take up this work of their own choice, albeit mostly out of financial need. The often relentless discussions about this are, in my opinion, not least shaped by personal moral value systems. Rarely are the voices of sex workers themselves heard, and even more rarely of clients.

Even after the introduction of the Prostitution Trade Ordinance and after the opening of the hustler boxes, the discussion has not calmed down. Some people don't think it's okay that the state spends around 800,000 Swiss francs a year on a prostitution booth. On the other hand, the primary aim of the boxes is to protect the population from the unpleasant effects of prostitution and to stop the ugly side effects that have arisen, especially on Sihlquai, such as the harassment of prostitutes by passers-by. When social work in the sex trade stands behind the sex workers as an advocate, it is always faced with a dilemma, namely, on the one hand to commit itself to the women and their situation and to demand good working conditions for them, while on the other hand knowing that perhaps this very commitment will lead to the women losing their jobs altogether.

Public and Private Social Work

The Zurich City Mission is a non-governmental organization, but subsidized by the city and canton of Zurich, the Reformed Church and with a mandate from Aidshilfe Schweiz. Thus we are independent, but nevertheless integrated into a diverse network of co-players. It is a special feature of Zurich that the state itself also runs projects in the area of sex work, such as the operation of the sex boxes. There is therefore a coexistence of private organisations and state-run services. The state grants subsidies to private organisations, which often encourages competition between the various organisations. Again and again the question arises, to what extent a participation in state bodies diminishes our independence and to what extent we can and must support a state attitude, closer to new regulations and laws. As in all social work, we have a dual mandate: clients here and funders there. The state's contribution to our work is about one-third of our budget. This means that we are obliged to disclose all our services to the state and to comply with guidelines on the full scope of our work. In contrast, the Reformed Church has not yet made any demands, either in terms of our attitude or our approach. Our work is certainly influenced by a diaconal attitude (especially by the sponsorship of the Reformed Church). Unlike other Christian organisations, we do not place this motivation in the foreground. Our work is not conditional and religion is only discussed if the client asks for it. The women who come to our counselling centre are usually in a more or less dire situation. We keep it with Bertolt Brecht, who said: First the food and then the morals. As a private organisation, we try to balance the above-mentioned tension, namely between good networking and cooperation with state bodies, other private organisations and church sponsors on the one hand, and an independent, personal attitude towards the sex workers that is justifiable for us on the other. In doing so, we are not least confronted with the social phenomenon of migration (Nowak, 2014).

3 Migration as a Core Burden

Basically, sex workers are in an exceptional situation because they have hardly any share in the democratic state. Due to their profession, participation in public life is hardly possible for them. They are stigmatized and usually do not even try to participate in public life. In prostitution, some of the rights of citizens are curtailed: the right to property, the right to a trial, the right to conclude valid contracts, freedom of speech and expression, and the right to move freely. For example, sex workers' contracts with their clients cannot be enforced in court, the regulations for a sex parlour are many times tighter than for any other trade, and sex workers who solicit clients in a place where this is prohibited are punished with expulsion.

However, this fundamental curtailment is very much exacerbated by the migration situation. This aspect of migration has often been left out of discussions and reflections or only mentioned in passing. Nowak (2014) in his book "Homo Transnationalis" has fundamentally drawn attention to this pressing issue for social work. In my experience, migration issues centrally shape the everyday lives of prostitutes and are far more burdensome than being a sex worker. Migration means: if they come from Europe, sex workers can work in Switzerland for three months a year. If they come from a third country, they are not allowed to work in Switzerland, unless they marry a Swiss or a man with a so-called C permit (longer stay in Switzerland). Dependence is the basic feeling and this sometimes forces them to marry. The consequences of migration are great poverty, no possibility of having one's own children with them, concealment of experiences of violence, no secure residence status and no social insurance. Problems arise that are more serious than problems resulting from their work as sex workers. In our work, we try to place this circumstance at the centre of our action strategy.

4 The Hot Topic of "Power"

In connection with prostitution, the issues of power, power structures and powerlessness on the one hand and the organisational structure on the other deserve special attention.

The Two Main Orientations of Any Social Work
First, I emphasize that the foundation of all social work is human rights. Central to these are:

- The right to liberty and security (Art. 3)
- Right to be recognised everywhere as having legal capacity (Art. 6)
- Equality before and protection by the law (Art. 7)
- Right to found a family (Art. 16)
- Right to free choice of occupation (Art. 23)
- Right to work equal pay for equal work without discrimination (Art. 23)
- Right to (social) security in the event of unemployment, sickness, invalidity, widowhood, old age or other loss of his or her means of subsistence through no fault of his or her own (Art. 25)
- Right to education (Art. 26)

Therefore, according to the United Nations guidelines, the first main orientation of all social work is to argue and struggle for social justice: "The emphasis that the social work profession places on basic human needs also determines its conviction that the universality of these needs and their satisfaction is not a matter of subjective choice or preference, but **an imperative of social justice.** Accordingly, social work is moving toward the conception of human and social rights as a second organizing principle of professional practice that complements the first organizing principle of needs orientation" (United Nations, 1994, p. 4, emphasis by author). The other main orientation is higher loyalty with concrete people before loyalty to an organizational size. "Social workers usually work as employees of the state or for large organizations, this has put many of them in difficult situations. The profession is beholden to both, the employer and the clientele. Because of the professional code, as well as the educational goals of social work colleges, **service to people** takes precedence **over loyalty to the organization**" (op. cit., p. 4). So what does this mean for social work specifically in the sex trade? After all, each of the aforementioned human rights is based on a legitimate and central need, which applies to sex workers just as it does to any other human being. However, very few sex workers (I am not talking about sex workers in the luxury segment here) know what their rights are, nor how they could get their rights. This is where one of our commitments in the counselling centre comes in.

Prostitutes' Experience of Power and Violence and Their Ambivalent Behaviour Towards Power

But there is a big obstacle – namely the serious insecurity in dealing with power. First, a definition in three steps from my teacher Silvia Staub-Bernasconi, who devoted herself to this topic again and again: "Power refers to rules or norms that prescribe how to distribute what is needed and desired by people and how this distribution is to be controlled. These rules may or may not be known, that is, power exists whether we think of it, whether we feel it, whether we think of it as central or incidental to our lives. And in any case, power regulates the relationship between at least two people in such a way that it results in a relationship of superiority and subordination or a relationship of control" (Staub-Bernasconi et al., 1996, p. 4). It follows that not only do prostitutes suffer from power, but that the issue of power also plays a central role in all social counselling. Firstly, regarding sex workers' experiences and handling of power: They have often experienced power as the exercise of violence and for some of them power is taken for granted until violence from men and power from the state. Silva Staub-Bernasconi (2012) already distinguishes between "limiting power" and "disabling power" in 1989:

Limiting power refers to those social rules that enable, legitimise (or allow) and support structures that are appropriate to human beings, i.e. structures that meet needs and promote learning.

Boundary rules ensure that no violence is used against people and things to enforce rules. Limitation rules and decision-making procedures are controllable by all and are therefore transparent. These rules apply to everyone. Limiting power enables human coexistence based on fair rules.

Disability power refers to those social rules that enable, legitimize (or allow) and support inhumane structures, i.e. structures that deny needs and restrict learning.

Disability rules use violence against people to enforce them, if necessary or systematically, and take resources that should be available to all only for themselves (appropriation). The self-imposed law of the strongest applies. Not all people are equal before the law. The rules are neither fair nor transparent.

The prostitution trade ordinance and various police laws, but also the fact that those working in the sex trade are mostly women and come from the poorer part of Europe or the world, show unmistakably that prostitutes in particular are at the mercy of the aforementioned obstructive power. This results in a highly ambivalent relationship of prostitutes to power. Almost everywhere the prostitutes smell a bad handling of power and yet they go on the defensive and "cower", albeit more or less "growling". This double attitude is sometimes or initially also evident in the social workers and their institution. At the same time they have learned to use their power. For example, by telling exactly the story that opens our heart or our wallet, or by cleverly using the dependence of the clients for pleasure satisfaction.

The Handling of Power to Be Learned by Social Workers

No less, I emphasize again in the context of Silvia Staub-Bernasconi, social workers in particular are at risk of not being able to deal well with power. As early as 1989, she spoke of a "trilemma" in dealing with power in social work, which consists of the following: "First, to criticize power and second, at the same time, to claim it for the clients and for oneself in order to realize certain demands, which means, third, that its bearers must learn to deal with their own and others' power... But this is precisely what makes them vulnerable again to criticism of power from their own and others' ranks" (Staub-Bernasconi, 1989, p. 4). I have made the experience that social workers in general and especially in the sex trade react incomprehensibly or one-sidedly when the topic of power is brought up. Often the attitude prevails: We social workers are ultimately helpless in the face of state power. In addition, they become small and sink into feelings of guilt when the suspicion arises that they themselves are possibly power-hungry or misuse their power. Especially social workers who work in the sex milieu for a long time are, in my opinion, in danger of identifying too much with their clientele and feeling as stigmatized and powerless as they do. This even goes so far that, for example, when a police officer appears, they immediately take a defensive or flight posture instead of confronting him or her in the attitude of a person of equal rank.

It is therefore indispensable for me that we social workers become much more aware of how we are entangled in the personal and collective power and powerlessness structures: when we ourselves use power clumsily or when we allow ourselves to be discouraged too quickly by state and social power. I find Silvia Staub-Bernasconi's proposed distinction between limiting and disabling power helpful in this regard. Social workers in the sex trade in particular are allowed to learn to demand power and to defend it, but without becoming power-hungry themselves or feeling powerless in the face of the state and society. They are allowed to become self-confident role models for their clients in dealing with power. This requires professional training and further education.

The topic of power and powerlessness also plays a role in the cooperation with other institutions that are not directly related to the milieu (e.g. institutions of the women's and peace movements). This cooperation is a counterforce to the not infrequently encountered nimbus: "Nobody understands us and we alone know about our clientele and the topic". I emphasize again: To understand the prostitutes in their world from "inside" is one thing, but it is just as important to make them understandable to others "outside" and in this way to get new allies.

5 The Variety of Methods in Social Work with Prostitutes: Illustrated by the Work of Isla Victoria

Mission Statement and Concept

The organisation Zürcher Stadtmission has a mission statement and the counselling centre Isla Victoria has a corresponding concept. The basis is human rights and thus the framework for the methodical work of our social workers. The Zurich City Mission recognizes sex work as a legal trade. It is important to create working conditions for this trade that are humane. The Zurich City Mission is committed to helping sex workers through its contact and counselling centre Isla Victoria. It respects the living and working situation of sex workers and takes their side on the following points:

- Isla Victoria works with respect and consideration for the dignity of women.
- Sex workers should be able to work in a self-determined manner.
- Pointing out perspectives for improving living and working conditions.
- Women affected by violence are supported and strengthened to stand up for their rights.
- Clients are asked to respect the sex worker and to value "fair and safe sex".
- Prevention of HIV/STI and general health care are offered.
- Women who want to leave sex work are supported in doing so.
- Social recognition of sex work is being actively addressed.
- The counselling centre works in a network with other institutions.

In everyday professional life this means for us that we see every woman in her dignity and that we see and promote her competences. To accept every woman as she is at the moment and to start working with her on this basis. To allow every woman the right to self-determination, to point out her duty of self-responsibility and to constantly try to realize these rights. And ultimately, to help them better understand their situation themselves. The offers of Isla Victoria are in this context:

- Advice in all situations
- General meeting place
- Meeting place for mothers and children (project "Tapis")
- Outreach work in the milieu (street work)

In the following I will briefly outline the most important methods for us and show how we apply them: social individual case assistance, street work as outreach work, social space orientation and community work. An important basic attitude is the so-called empowerment. Methods are theoretical plans of action to tackle a problem. However, reality rarely allows methods to be applied in their pure form. Therefore, every method has its limits, so that, among other things, it must always be asked whether and how it might suit the target group, whether it makes sense within the institution and whether its effect can be verified. We have made the experience that all main methods are indispensable, even if a team is very small. Delegating one of the main methods to other agencies does not seem possible and certainly not sensible from our point of view. This corresponds to the basic "all responsibility" in social work (Galuske, 2007, p. 36). For example, we see a social worker in outreach work (street work): she encounters a crisis situation and has to apply individual case assistance. A little later she meets with a colleague from another institution in order to take the next step within the framework of community work.

Problem-Solving Individual Case Assistance

Individual case assistance focuses on one individual. The goal of the assistance process is in all cases the satisfaction of the client's basic needs. We represent a problem-solving approach here. In the process of life, people repeatedly encounter problems they cannot solve and need support to solve them. The supporters must therefore motivate and enable those affected to discover and accept resources. Foreign sex workers are rarely familiar with the system of counselling services in Switzerland. Although they are informed of the counselling centres when they apply for a licence to work, they do not really understand this information at first and forget about possible counselling services. In the meantime, however, we can build on a well-functioning information system among the sex workers and refer the prostitutes to each other, for example, to our meeting place and the lunch served there. The women come to lunch, where the social workers are also present and available for short consultations. So often the individual case help with us begins with the handing out of a meal and with the obtaining of condoms. This is followed by a short, perhaps even non-binding question about a room or a place to go. However, it also happens time and again that women with their entire disordered and tragic life story virtually walk in the door. And then all the issues of social work can become clear, such as financial hardship, violence, homelessness, health problems, pregnancies, residence problems, etc. Not in every case it is immediately clear whether a classical help process can be started. Our offer is not bound to any conditions. Some women may only want to have the acute emergency remedied without solving the underlying problems. Nevertheless, it is important to quickly get a picture of the woman's situation: What is her concern, what is the most pressing problem, and are we the right place to address the problem? A case history must be taken in such a way that the woman feels safe in the interview, especially since many sex workers are familiar with interviews with police officers and quickly react in a closed manner. For us, individual case assistance often means providing a small amount of emergency financial aid. But it is to be asked how far this makes sense and is purposeful. Sex workers have a special relationship to money. Sometimes they earn a lot of money and spend it immediately (or send it home) and sometimes they earn no money at all and cannot fall back on savings. They live entirely in the moment and planning for the future is rare. But if a woman can no longer charge her mobile phone, she may not find any customers and no longer earn any money. A few francs can be very important. A sex worker who has no money cannot immediately apply for state social welfare. Exceptions are women who have been in Switzerland for a long time and who no longer want to work in the sex trade. Any person who does not have a so-called unlimited residence permit will be deported after about two years of receiving social assistance. So it is important to weigh up in each individual case whether a small amount of support can really remedy the hardship.

Many sex workers who come to counselling have grown up without education and speak little or no German. For this reason, the conversation must be very simple, without psychological vocabulary, and in any case trust-building and accepting of the situation. Mrs. X comes to counselling because she wants to separate from her husband and reports that he is often aggressive and drinks. The woman only has a so-called B permit, which means that in case of divorce she would lose her right of residence in Switzerland and would have to return to her country of origin. The social worker shows her the advantages and disadvantages of separation and divorce. She has to find out the woman's room for manoeuvre, however small it may be. Rarely do the women want to go back to their home country. Should we support them in staying with an aggressive husband? Here, many structural obstacles or the aforementioned disabling power become apparent. Social work in this environment often has to look for the solution that does the least damage. We notice again and again how on the one hand we want to represent the ethical basic attitudes, but on the other hand the possibilities of action are incredibly limited.

In contrast to state social assistance, we have no control functions. At most, we can stop our assistance if the woman is no longer expected to participate in the solution process. However, we have to take into account the fact that very few women have ever experienced what such a process of participation can mean. On the other hand, sex workers in particular have incredibly high competencies when it comes to surviving in everyday life. Scepticism and interest change constantly. Social work with sex workers has to accept the fact that it is never entirely clear how long a process can last and it has little influence on the further course. Perhaps the woman changes her place of work and moves away, or her residence permit has expired, or she no longer wants counselling.

Empowerment as a Desirable Basic Attitude

Empowerment is mentioned in almost every mission statement and concept for counselling sex workers. Strictly speaking, empowerment is not a method, but a basic attitude. Empowerment wants to move away from a deficit-oriented view towards a promotion of strengths. "Empowerment means the process within which people feel encouraged to take their own affairs into their own hands, to discover and take seriously their own strengths and competencies, and to learn to appreciate the value of self-generated solutions" (Keupp 1996, p. 164, cited in: Galuske, 2007, p. 262). At the same time, it is an opportunity to join forces with people who are equally affected. Empowerment always takes place on three levels: individually, in groups and on a structural level. The individual level is included in individual case assistance.

Empowerment is also something that social workers working with sex workers want to enable precisely because, in their experience, these women have many strengths. But this is a mistake, because in my opinion there is a big gap between knowing about the strengths and making them effective. And some strengths are only good for very specific limited moments, not for problem solving. It is one of the special characteristics of sex workers that while they are often victims of "exploitation situations", they just as often experience situations where they are the power holders. On an individual level, they experience almost daily that they also have strengths and, at least momentarily, power, especially power over the satisfaction of the johns. They are the ones who enable their (supposed?) satisfaction, they determine who can 'shop' with them and they listen. But they also determine what they themselves feel, or don't want to feel. They often give the individual man the feeling that he of all people is loved by them and that nothing is pretended according to the motto "I am the great exception". The question is just how this strength can also be used to cope with problems. Unfortunately, they also live out this strength negatively, for example by using police officers to gain advantages. Sometimes this strategy succeeds, but not infrequently the women are also reported.

In addition, there is the considerable difficulty that these empowerment processes take a lot of time, but the women often only work in one place for a short time. Empowerment of the sex workers would result in a joint standing up and speaking out, but the women often shy away from this. So far we have only managed once to get women to sign a resolution.

Rudimentary Approaches to Group Work

In order to nevertheless address the level of the group, we have engaged a sex worker who produces creative works with the women one afternoon. In this offer, painting, handicrafts and writing are done, whereby the social worker stays in the background. The work is exhibited in public, albeit on a small scale. The aim of this activity is, on the one hand, to give the sex workers the opportunity or experience that they can also do other things than sex work. On the other hand, it is important that they can experience this in a group. These are the very first small building blocks of the goal to experience one's own strengths. In our sub-project, the social workers work with prostitutes and their young children. It is a kind of meeting place with professional child care. The focus here is on the one hand the effort within the empowerment process to address the sex workers as mothers and additionally to promote their being together with women in a similar situation. However, it quickly became apparent that instead of empowerment, practical life support services were initially necessary.

Perhaps empowerment is also mentioned so often because sex workers are so far removed from it. For "empowerment [is] largely shaped by an image of the autonomous and successful individual who is able to pursue his or her interests and ideas together with others" (Stark 1996, p. 76, cited in: Galuske, 2007, p. 266). However, we must soberly acknowledge: Sex workers hardly have these prerequisites.

Streetwork as Outreach Work

A second important pillar for us besides the meeting place is the outreach work in the milieu. This is where initial contacts are made, where the scene is observed and where first aid is provided.[2] "Street work is a form of contact in the sense of outreach work. Street workers do not (only) work in the rooms of an institution, but (also) go into the immediate living environment of their target group by visiting their informal meeting places" (Gref 1995, p. 13, cited in: Galuske, 2007, p. 268). This ranges from street corners to shopping arcades to private rooms and apartments. Although outreach work was only named as a method in the 1970s, the Zurich City Mission already carried out this form in the 1950s, at that time as the so-called "Midnight Mission", on the street. We have retained this method up to the present day, albeit with different objectives today.

In principle, the street work method addresses more primarily problem-ridden social spaces and subcultures and aims at changing or at least controlling the behaviour of the (delinquent) groups. In the prostitution scene, establishing contact is the first goal. Sex workers have more difficult access to counselling services. Due to their remoteness from education, they are not used to taking written messages. Many do not have internet access. Their working hours are often such that they have little time to visit the counselling centres. The latter concerns especially women who live further away from the city. Outreach work requires that the social workers know the relevant places of work, such as cabarets, salons, clubs, bars, apartments, etc. They need to know when they can enter where, because not every club owner or barmaid is pleased when social workers keep customers up or away from the business. They need to know the different scenes and move around these places in a way that does not act as competitors with the women or as "enemies" of the owners. Outreach work must not be perceived as controlling. Here, too, the aim is to create trust, to give initial information about the counselling possibilities and, if necessary, to provide first aid.

The double mandate of any social work (loyalty to the client – representative of an ethical organisation) also plays a role here, albeit with a different emphasis. We bring the help closer to the everyday life of the clients and at the same time we become accomplices of illegal situations or crimes. It is always necessary to carefully weigh up in each individual case which steps are in the best interests of the woman and which we can justify in the light of our mandate and ultimately our conscience. The more the scene retreats into illegality and the more women move into the non-metropolitan area of the canton because of the laws of the city, the more difficult the outreach work becomes. For example, until two years ago we were able to approach women in the red light milieu on the street and distribute

[2] Michael Galuske points out that there are many project reports and concept papers on this area, but still "no satisfactory introduction" (Galuske, 2010, p. 128). So here we have to very consciously keep trying out new and creative steps.

condoms. That is no longer possible today, because it is absurd to distribute condoms in a place where the women are not allowed to prostitute themselves. We have made the experience that many women first seek help for health problems. If they find trust through this, they then also turn to the counselling centre for social or legal problems. Thus, we provide outreach work with trained nurses who can keep an eye on minor health problems (e.g. measuring blood pressure, initial clarifications regarding STI).

The social worker in outreach work takes a holistic approach. She has to be available for almost all problems, has to provide crisis intervention, carry out accompaniment, organise housing, etc. Our employees as street workers are the antennae of our counselling centre – they are the first to notice when the scene changes, they hear when pubs are to be closed and they know which pubs are good places to work and which are not so good. As employees of the counselling centre they also work on the premises of the counselling centre and advise women they know from their outreach work.

Our special mandate from the Swiss Aids Federation is HIV/STI prevention among female sex workers (FSW). We also carry out this task primarily through outreach work. Condoms are often a door opener for entry into the establishments and are gladly accepted as gifts by the women. If the street worker finds that an intervention with the employer is needed, she refers the case – exceptionally – to another counselling centre for intervention. If we were to carry out such interventions ourselves, access to the establishment would be blocked. It is a matter of helping the women to assert their rights, such as demanding wage or insurance payments from the employer.

There is a model for prevention in Switzerland through so-called "mediators". These are former sex workers from different ethnic groups who visit sex workers from their ethnic group and hand out condoms as their main task. This model was tested in our counselling centre before my time, but in the meantime we have distanced ourselves from it. This is because we do not think that handing out condoms alone is effective enough. In addition, former sex workers are often part of the scene and not really independent and also not sufficiently trained for more complex questions.

6 Social Space Orientation: Illustrated by a Case Study

Definition and the Five Method Principles

The social space of prostitution is often public space or it borders on public space. In this context, social work in the social space has the following goal: "Social space-oriented work does not aim at the 'improvement' of people, at the targeted change of their living habits or educational intervention regarding their communication styles, but at concrete improvement of the living conditions of the residential population in a residential quarter with the active participation of the people concerned" (Hinte and Kreft 2005, p. 870, cited in: Galuske, 2007, p. 277). People take offence at the existence of the trade from the public space, even if it takes place in the private space. Often the sex worker's place of work equals place of residence. Not only because of prostitution, but because petty crime, drug dealing, pleasure activities and migrants congregate around the milieu. this public space is very much policed. At the same time it is a fascination, tourist attraction and market place. Social space oriented work is about improving the quality of life of the people and here in particular of the sex workers. Central is of importance what the affected person wants and how the participation of the affected person can look like. We know too well that help that only gives and does not require any personal contribution supports dependency and undermines dignity.

The Essen School for Social Space Orientation names five "methodological principles" (Früchtel et al., 2007, p. 40):

1. Orientation towards the interests and the will
2. Support for self-initiatives and self-help
3. Use of the resources of the people and the social space
4. Cross-target and cross-sectoral approach
5. Cooperation and coordination

How can sex workers be won over to participate? How can their will and interest be found? The following description of the life story of Antonia from Bulgaria serves as an illustration:

Antonia is found by our street worker in the milieu. She no longer has a place to live because she has been living in her boyfriend's room without the landlord's knowledge. She is staying illegally in Switzerland. The boyfriend is thirty years older than her and a welfare recipient from Germany. Antonia is eight months pregnant. The social worker is able to accommodate Antonia in a room at Café Yucca, another department of our Zurich City Mission, until a month after the birth. The older German friend Willi promises to look for an apartment in Germany, as Antonia has no prospects of being able to stay in Switzerland. Antonia is registered as a tourist in Zurich. She thus frequents our Café Yucca for people in difficult circumstances and meets Hans there. The pregnancy is taken care of by the social worker and nurse and health insurance is taken out. Hans strongly courts the pregnant Antonia. The staff at Café Yucca know Hans and know that he always has women's stories going on. A woman who had experienced a story with Hans tries to warn Antonia. There's also a rumour going around the café that Hans is suspected of child abuse. He has a 15 year old daughter. Hans is in poor health and is about to have an operation on his arm. A week before the due date, Antonia and Hans open up to the social worker that Antonia wants to move out of the Stadtmission and move in with Hans. They have known each other for about three weeks. That same evening the child is born. Hans accompanied her at the birth and lays claim to the child. Antonia expresses great disappointment that the German father Willi does not want to see the child at all. However, she had told Willi that she now had a Swiss boyfriend who wanted to marry her and so she could get a residence permit. The impression was created that Antonia and Willi's relationship was over.

As the leader of the Zurich City Mission, I organized a helpers' conference in the hospital. The ward nurse, the nursing specialist, a social worker from the hospital's social services, the social worker from Isla Victoria and I as head of the Zurich City Mission were present. The conversation revealed that everyone involved had great reservations about letting the woman go home with the child. There was talk about reporting the case to the Youth Welfare Office for endangering the welfare of the child. The final decision on whether or not to take action rested with the hospital's social counselling department. The social worker wanted to involve her boss. In particular, the ward nurse did not want to let the mother go home with the child. Already in the hospital, Antonia reacted to the child's crying in a very angry, indignant to completely helpless manner. The next day the mother was informed of the decision that her child would have to stay in hospital. She received this decision with great anger and aggression and threatened that she would travel to Bulgaria with the child. Hans, on the other hand, accepted the decision with equanimity and thought that they would come to an arrangement.

Again one day later the hospital decided – surely for cost reasons – that mother and child would be discharged home after all, especially since a midwife had been found who could visit daily. The social service of the hospital made a child welfare endangerment report. In the meantime, discussions took place with the child's father. He was deeply distressed that Antonia did not want to know anything more about him and he was afraid of Hans. Later it turned out that Antonia was playing a double game and still wanted to be with Willi. The Zurich Youth Welfare Office passed the case on to the suburban municipality where the couple lived. Antonia and Hans tried to get support and help in various places, e.g. for the translation of Antonia's papers, for the preparation of a subletting contract and for other needed money. The social worker has withdrawn from the case as the Youth Welfare Office is now in charge. Also the staff of Café Yucca, where the two often drop in and show off their child, no longer give financial support. The hospital contacted me as the head of the city mission because Antonia did not come with the child for a follow-up examination. I then called the social worker from the youth welfare office and also mentioned the suspicion of child abuse of Hans. This suspicion was not passed on by the hospital to the youth welfare office. The social worker from the youth welfare office was very concerned about the situation. The story is probably not over yet and it shows how involved and complex the concrete social work involved can be.

The Five Method Principles in the Case History

- Orientation towards the will and interest of the client

We had the impression that the client's primary goal was to stay in Switzerland, especially since she was living on the streets. Measured against this goal, the choice to take her friend Hans from Switzerland as a partner was certainly the right one. She wanted to change her uncertain status and used the resources available to her to do so. Our value system makes us question this relationship and we ask about the client's motivations. The focus here, however, is not on the actions of the adults, but must always be on the well-being of the child.

- Support for initiative and self-help

Hans in particular has developed his own initiative. He tried very hard to get children's clothes and also to get the legal matters in order. But Antonia's own initiative diminished as a result, or did not arise at all, signalling the message "please do this for me". Hans was willing to do something for her, but not in a position to do something with her and thus promote her self-help. However, it would be the task of any further social work with the couple to promote this.

- Use of human resources and social space

No social space existed in this case history for longer than three weeks. The infrastructure of Café Yucca (including accommodation) was used because Antonia could eat and socialise there. Whether we were able to uncover Antonia's resources in this short time is questionable. During the pregnancy she showed herself to be very helpless. We had tried to use the resources of the hospital social space. It would certainly have made sense to leave the child in the hospital for a few more days and observe how the relationship develops on the part of the mother. This plan failed due to financial concerns, as it turned out that no one would pay for the child's stay in hospital. Basically, it should be noted here that our social workers know the social space very well. The question remains, however, whether they can use this knowledge sufficiently for their clients.

- Cross-target and cross-sectoral approach

Antonia was a sex worker and met her first boyfriend Willi during her work. Mostly, however, she was seen as a pregnant woman and woman in childbed and treated as such. Hans, a guest at Café Yucca for marginalised people, receives a disability pension and comes from a different target group. Different sectors had to work together in this case: Café Yucca for marginalised people, which also rented the room to Antonia, the Isla Victoria counselling centre, the hospital's nursing staff, the hospital's social services, the youth welfare office and the social services of the suburban municipality. Finally, two other private organizations were involved, in which Hans also frequents. In order to bring the living situation into a safe and responsible network, the migration office, the maternity counselling service in the locality and Hans' counsel would now have to be involved. One question in this context was how to integrate the couple into non-governmental and non-deficit-oriented groups?

It remains open whether Hans will get a marriage license, whether he wants to continue to marry Antonia and whether a safe solution can be found for the child. For our counselling centre, this means continuing to be vigilant when inquiries come from Antonia or from Hans. But it also means getting involved when it comes to migration policy, when sex workers who are mothers are stigmatised and when the welfare of the child is in danger. And this even if we have to put the client's will second.

- Cooperation and coordination

A lot has already been said about cooperation and coordination in this case. In principle, we try to achieve cooperation with other professionals. However, I would like to mention the difficulties of such cooperation. The improvement of the quality of life of our target group can seldom be clearly determined. Basically, we find that the rent for a room that the prostitutes rent is extremely high. Is it now right to report this landlord with the certainty that the woman will then have no room at all? Is this an improvement of the quality of life? We have decided on a division of labour. The social workers work with the women and look for the solution that the woman prefers, even if we find it unfair. I, as head of the Zurich City Mission, on the other hand, work in political committees and with the help of media work to improve conditions for the women. But even at this level the question is asked:

Should the police give permission for a salon that charges more for a room than seems right to us? What happens then? There is one less salon and the women lose their opportunity to work. Finding resources would probably mean renting out rooms ourselves! We cooperate with a commission of experts which accompanies the implementation of the prostitution trade regulation. As a non-governmental organisation (NGO) we have to point out again and again the points which are difficult for the sex workers and we have to criticise e.g. the constant controls. The participation of the citizens often takes place in the form of complaints, which the police have to investigate. All those citizens who have no problem with sex work are hardly heard. It is also important to consider unusual forms of cooperation. For example, we were able to cooperate with the city theatre because they put on a play about prostitution (by Elfriede Jelinek "Über Tiere") or we found contact with a circle of women writers who want to approach the subject. Critically, we have to ask what happens to those clients who do not let themselves be activated, who do not express a will. Are these people left to the regulatory powers?

Community Work: A Still Open Field

Community work is closely related to social space. In Zurich, the traditional red-light district "Langstrasse" is increasingly being converted into an in-quarter with chic apartments and high rents. Prostitution is being pushed out of this quarter more and more. At the same time, the whole quarter is becoming an event space where young people hold their parties on the street at weekends. In view of a presentation given by Christiane Howe, Humboldt University Berlin, in Zurich,[3] it became clear to us as listeners that we actually have no idea what the population wants. Does the population of Zurich want to have a district in Kreis 4 (the centre of prostitution and prostitutes) that is dead and empty in the evening? Perhaps, on the one hand, we have become too accustomed to the fact that sex workers fundamentally shy away from participating in the community. They don't want to show themselves, they don't know the language and they have conflicting interests. On the other hand, we forget to think about the groups of residents and neighbourhoods affected.

We have probably done the same as the municipal community work. Even the urban community work of district 4 – precisely the focal point of prostitution – has never launched a project that focused on the issue of prostitution. Here lies another

[3] Christiane Howe (from the Institute for European Ethnology at the Humboldt University of Berlin): Nachbarschaften und Straßen-Prostitution. Konfliktlinien und Lösungsansätze - Am Beispiel Kurfürstenstraße in Berlin-Schöneberg. Lecture in the context of the Expert Commission on Prostitution of the City of Zurich 22 August 2014.

opportunity that we have not taken advantage of, and we are beginning to wonder how we might do a better job of bringing our concerns to the attention of local political bodies and administrators. Prostitution is a part of every major city and community. Even tolerant people, however, I hear say, "Yeah, sure, but please not where I live or work." It takes community work to be able to change such opinions and attitudes. If we approach community work more actively and decisively from our side, then it cannot be ruled out that the sex workers can also be more activated in this way.

7 Challenges and Consequences for Social Work

I see the following conditions, concentrated on Switzerland, as particular challenges for social work:

- Competition among women (and in some cases also among ethnic groups) in the trade is permanently very high.
- Some women are controlled by pimps.
- The johns are anonymous and hardly show themselves.
- No less is the anonymity of women part of their business and life.
- The social space is well known, but hardly accessible to non-insiders.
- Residency status is not always legal.
- There is – due to migration – only a sporadic and short duration of stay.
- Again and again there are conflicts with the law.
- The effectiveness of social work with prostitutes has been difficult to demonstrate.

Competition among women tempts them to work little or not at all in groups. Valuable mutual help opportunities are lost in the process. Social workers in prostitution have to "jump" all the time. For example, they sometimes work simultaneously with women who want a late-term abortion and with women who want a child so that they can get better recognition. They sometimes have to be able to navigate the not-quite-legal terrain that prostitutes enter. They are bearers of knowledge that most people are immensely interested in and which they are nevertheless not allowed to divulge. In no other target group of social work does the system of pimps exist. Even if far fewer women are dependent on pimps than is commonly assumed, this phenomenon is still formative. An equally difficult area is the johns. From my perspective, they could be great resource persons, but they remain mostly anonymous and invisible. Women, however, consistently tell how johns take care of them and deal with them well, though of course motivations would need to be clarified. Since time immemorial, poverty has also (but not always alone) driven women into prostitution. Today, prostitution can only be considered from the political-social situation in the whole of Europe. The goal would be international networks and transnational projects that stand up for the rights of sex workers. Last but not least, the impact of social work with sex workers usually remains unverifiable. Even if a social worker can show an impact, there is still no proven system of how this could be qualitatively evaluated The only indicators so far are quantitative figures such as contacts, visitors and condoms distributed.

The following consequences for professional social work can now be drawn from what has been said so far.

1. Professional training

From all this it follows, firstly, that social workers in prostitution must have a great many skills. They must have good knowledge of human nature, be able to deal with psychological difficulties and at the same time with the authorities and the police, and at the same time have a high level of self-reflection. The intensive proximity to the milieu can, as already described, especially if it lasts for years, tempt the social workers to behave like their clients, i.e. to feel devalued, to always have to protect themselves and to remain in the familiar circle. Here, a professional specified training is needed, in which it is also checked whether a social worker has the necessary prerequisites for this multifaceted work. Only then could they become role models for their clients – not least in dealing with power. They must also continue to have the means to document the effectiveness of their work more precisely and finally have the courage to consciously address the three further consequences that now follow.

2. Networking with other professional disciplines

Secondly, there is a need for networking with other professional disciplines. A good network of reliable professionals and institutions is needed, such as lawyers, doctors, civil servants, social welfare offices, migration offices, police, specialised offices for trafficking in women, etc. All of these need to know about prostitution and the situation of women in prostitution so that they can act accordingly. All of these need to know about prostitution and about the situation of women in prostitution so that they can act accordingly. Our experience repeatedly points to the need to set up a round table on sex work in Zurich. In it, all those affected should have their say and joint solutions should be worked out.

3. Special handling of the methods

Thirdly, the high challenges also require a special approach to the methods. If we assume that prostitution is a trade like others, if we want to treat sex workers as people like you and me, then we must also apply the general methods for this social work, but in a modified and above all conscious way. As in any social work, it is necessary to test them with many target groups, to constantly review them, to adapt them to the current situation and to acknowledge their limitations in each case. The work with sex workers e.g. in Sweden on the one hand and in Switzerland on the other hand is quite different due to the political-legal situation. But I believe that the basics of human rights, the question of power and the development of resources in small steps are non-negotiable.

4. The need for private social work – in the struggle with the state authorities

Fourthly, the special challenges require the structures and services of private social work. State social work's mandate is focused on the legal basis and its enforcement with the help of the police. Private social work is closer to the people and can, in my opinion, better fulfil the dual mandate of social work (respect for the law and society on the one hand – representation of the rights and needs of the clients on the other).

For example, the situation and the development of the last few years in Zurich shows that the state as the guardian of order and protector of the population can only to a limited extent also perceive the concerns of the sex workers. Of course, the state also has a dual mandate: to protect citizens, including public space, and sex workers. But its focus is on protecting citizens and public space. Thus, the model of the hustle boxes is indeed an attempt to control the hustle by the state and at the same time to care for the women in a social work way. But women who are afraid of state institutions (for whatever reason), or women who want or need to escape all control, naturally do not get involved with cantonal offers. Here, private or non-governmental organisations (NGOs) are absolutely necessary.

Foreign women in particular, who have not experienced the state and the police in their home country as "friends and helpers", are better able to build up trust in NGOs. Although the NGOs are also bound by the state's guidelines, their advocacy work is not disturbed by the presence of the police. Moreover, NGOs can often act quickly and unbureaucratically in changing situations. In contrast to state organisations (which are usually very large), small NGOs can implement a participatory management style that delegates as much decision-making authority as possible to the social work on the ground. This gives them the chance to learn from the sex workers as well and to make this knowledge useful for their work. However, the NGOs must engage in discussions with the state authorities and champion the cause of their clients, knowing full well that the latter do not understand and accept the NGOs' point of view in every case. All the more reason for NGOs and their social workers to be willing to advocate politically for their goals and values. The NGOs and their social workers must be self-confident, open and clearly committed to women's rights and declare this, not least to the representatives of the state authorities. They are the bodies that live a counterforce to the social devaluation of prostitution and prostitutes and are therefore allowed to publicly declare their interests. In this way, the professionalism of this work can become effective in a larger context.

Finally, one more of my visions: From a Swiss perspective, it would be (more than) sensible to set up an independent specialist body that can ensure political work, professional development and cooperation with the universities of applied sciences.

References

Agustín, L. (2005). Migrants in the mistress's house: Other voices in the "trafficking" debatte. *Social Politics, 12*, 96–117.

Bugnon, G., Chiementi, M., & unter Mitwirkung von Chiquet, L. (2009). *Der Sexmarkt in der Schweiz. Kenntnisstand, Best Practices und Empfehlungen* (Teil 1 – Literaturbericht). Universität Genf.

Dücker, E. von (2005). *Sexarbeit. Prostitution – Lebenswelten und Mythen*. Anlässlich der gleichnamigen Ausstellung im Museum der Arbeit, Hamburg. Bremen: Edition Temmen.

Entresol. (2014). Bezahltes Begehren. Ein Workshop zum Problem der Sexarbeit. http://entresol.ch/intern/veranstaltung/bezahltes-begehren-workshop-zum-problem-der-sexarbeit . Zugegriffen: 2. Jan. 2015.

Frauen MA57 – Frauenabteilung der Stadt Wien. (2013). *Internationale Vergleichende Studie zu Prostitutionspolitiken: Niederlande, Österreich [Exkurs Schweden]*. Studienautorinnen: Wagenaar, H.; Altink, S.; Amesberger. H.; Redaktion: Gebhart, M. Übersetzung: Szabo, S., Wien: Stadt Wien.

Früchtel, F., Budde, W., & Cyprian, G. (2007). *Sozialer Raum und Soziale Arbeit. Textbook: Theoretische Grundlagen* (2. Aufl.). VS Verlag für Sozialwissenschaften.

Galuske, M. (2007). *Methoden der Sozialen Arbeit. Eine Einführung* (7., ergänzte Auflage). Juventa.

Galuske, M. (2010). Straßensozialarbeit. In D. Kreft & C. W. Müller (Hrsg.), *Methodenlehre in der Sozialen Arbeit* (S. 126–128). Ernst Reinhard.

Gemeinderat der Stadt Zürich. (2012). *Prostitutionsgewerbeverordnung. Auszug aus dem Beschlussprotokoll – 114. Ratssitzung vom 7. März 2012*. Stadt Zürich, Zürich.

Nowak, J. (2014). *Homo transnationalis. Menschenhandel, Menschenrechte und soziale Arbeit*. Barbara Budrich.

Staub-Bernasconi, S. (1989). Macht: Altes Thema der sozialen Arbeit – neues Thema der Frauenbewegung. Versuche im Unterscheiden zwischen Behinderungs- und *Begrenzungsmacht. Sozialarbeit, 3*, 2–24.

Staub-Bernasconi, S. (2012). Macht und ihre verschiedenen Formen. http://www.kshm.at/fileadmin/unser_upload/PDFs/Macht.pdf . Zugegriffen: 8. Sept. 2014.

Staub-Bernasconi, S., Heiner, M., Meinhold, M., & Spiegel, H. (1996). *Methodisches Handeln in der Sozialen Arbeit*. Lambertus.

Ulrich, A. (1985). *Bordelle, Strassendirnen und bürgerliche Sittlichkeit in der Belle Epoque. Eine sozialgeschichtliche Studie der Prostitution am Beispiel der Stadt Zürich* (Mitteilung der Antiquarischen Gesellschaft in Zürich, Bd. 52, Heft 3). Schulthess AG, Zürich.

United Nations. (1994). *Human rights and Sozial work. A manual for schools of social work and the social work profession* (Series no. 1). Center for Human Rights Geneva.

Zürcher Stadtmission. (2011). *Die Lebenssituation von Sexarbeiterinnen in der Stadt Zürich und Konsequenzen für "Isla Victoria"*. Erstellt von Karin Meierhofer und Lukas Leutenegger von der Schiess Unternehmensberatung.

Regula Rother is the head of the Zurich City Mission. With the offer "Isla Victoria", a contact and counselling centre was created especially for women in prostitution.

Career Entry in the Field of Female Prostitution: Challenges in the Context of Social Work Counselling

Julia Kempl

1 Introduction

Although social work in the field of prostitution – especially female prostitution – has seen an increase in counselling services and new developments in recent years, it still belongs to a small and manageable area compared to other fields of social work. In many large German cities, there are already counselling centres for women who engage in prostitution. Depending on the location, there is therefore a greater or lesser awareness of and interest in this topic among the population. This also applies to students, especially of course to those who can find a counselling centre in the immediate vicinity of their place of study. With regard to internships and scientific work up to bachelor's and master's theses, students show an increased interest in getting to know this field of counselling more closely and researching it scientifically within their studies.

Despite this growing interest, prostitution and social work are hardly a topic in university education. The content and legal foundations of this field of action have not yet been taught during studies, or only to a limited extent. German-language specialist literature that could have concrete relevance to social work is also hard to find at present. For social workers who would later like to work professionally in

J. Kempl (✉)
Mannheim, Germany
e-mail: julia.kempl@diakonie.ekiba.de

M. Albert, J. Wege (eds.), *Social Work and Prostitution*,
https://doi.org/10.1007/978-3-658-37761-8_9

this field of action, there are therefore hardly any possibilities to bring along practical previous experience from this area. In addition, every counselling centre is confronted with different forms and problems of prostitution and therefore has to work with city-specific offers and concepts. The approach of a counselling centre is completely different in a city where there is more prostitution in apartments or on the street than in a city where there is more of a concentrated brothel street. Therefore, comparisons of the counselling centres with regard to their concepts, approaches and offers are by no means congruent nationwide. Every social worker who wants to work in the field of prostitution should therefore be aware of her personal motivation and expectations for working with this target group. The manifestations of prostitution are manifold and therefore entail the necessity of a differentiated and reflected professional attitude in everyday professional life, which must always take into account the individual case. Starting a career in the social work field of prostitution entails a complexity that should not be underestimated and that differs significantly from other areas of social work, where it is often possible to fall back on long-standing standardised procedures.

In this article, the complexity and the associated challenges of a professional entry from the perspective of social work in the field of action of female prostitution will be presented and examined. The question of the necessity of professional experience from other areas of social work, which is then applied and expanded in the field of action of prostitution, will also be addressed in this article.

2 Access to Women

Every counselling centre in Germany that provides services and support for women in prostitution finds city-specific conditions and manifestations of prostitution. From street prostitution to brothels or flats with appointments – to name just a few examples – social workers have to adapt to the circumstances of the local milieu. This inevitably entails different concepts and approaches in terms of access to the target group. In order to be able to inform women locally about the offer and to reach them at all, a sensitive consideration of the women's working and living environment is required. A possible form of access is shown by most counselling centres through the method of action in the form of street work. In the sense of Galuske (2011, p. 290 f.) there is already at this point the need for a lifeworld-oriented social work, which thinks and acts more than just purely institution-related. It enters a problematic social space in order to come into contact with a target group which, due to stigmatisation, tends to avoid existing institutional counselling services. On the one hand, it is often the case that the women concerned are unaware of existing services, and on the other hand, they are very reluctant to seek out institutions that are still unknown to them, as most women outside the milieu often have to contend with prejudices. The personal appearance and actions of the social workers within the women's social space are therefore decisive for successful communication and possible access. Only a sensitive and situation-oriented approach sets the basic prerequisites for a woman from the milieu to later contact the counselling centre on her own. It is helpful if women have already seen the counsellors through street work activities and can thus get a first impression of who they are dealing with. This can later be the decisive step for a prostitute to actually dare to go to a counselling centre that is still unknown to her. It should also be taken into account that it makes a decisive difference in the initial contact between the counsellor and a woman whether this takes place within the milieu or the working environment of the woman or in the rooms of the counselling centre. A special feature of the counselling work in this field of action is that only female social workers are active in the counselling contexts of female prostitution. There is also a clear gender division in the target group, since services for female and male prostitutes are usually clearly separated from each other.

While in other counselling centres the responsibilities are usually limited to one problem area of the client (e.g. in educational counselling), in the context of prostitution counselling there are many different areas that are taken up and dealt with in the counselling process, as they are directly connected to and dependent on each other. The counselling work therefore requires social workers to take a multi-perspective approach to the individual problem areas, which can differ greatly from person to person. A successful counselling process requires carefully thought-out and, if possible, ideal institutional framework conditions. The location and accessibility of the counselling centre are crucial for successful access to the target group. The atmosphere of the facility, the appearance of the counsellor and the equipment in the counselling room set the course for the counselling process. In order to make this process as successful as possible, a starting situation appropriate to the target group is needed. The various counselling contexts are as different as the women themselves: Some women first make contact by using the low-threshold services, such as open meetings, clothing distribution or medical consultations. For the counsellors, this is an opportunity to offer further contact and counselling services. By making use of these low-threshold services, many women gradually develop trust in the professionals, as they are available as contact persons due to their constant presence. This enables many women to decide to engage in a long-term counselling process.

Some women come to a counselling centre specifically from the beginning, have concrete concerns and express the desire for change or immediate action.

Here the counsellors are asked to point out their own limits and possibilities and to communicate these to the other person. Concerns and wishes that cannot be satisfied often show themselves directly through various reactions on the part of the women, such as disappointment or annoyance. The feelings that then become clear in the conversation must be professionally caught and addressed by the specialist. It is also important to convey to the women that social work boundaries do not imply personal rejection and are therefore not a matter of sympathy or antipathy. This is important in order not to negatively influence the relationship between client and counsellor.

As soon as it becomes clear that one is entering into a longer-term counselling process, it is of central importance to address and clarify the woman's self-motivation. In this way, it becomes clear from the beginning who is taking on which tasks in this process and can make expectations transparent. This means – as in other counselling contexts – creating an awareness of personal responsibility in the client. Particularly in the case of women who work in the field of prostitution, it is more common for them to experience themselves as externally controlled due to different dependency relationships in their working environment. In this respect, one of the main goals of counselling is often to help women regain their ability to act and their self-efficacy with regard to their own areas of life and ideas.

3 Competence-Oriented Challenges for the Counselling Specialist

In the counselling context of female prostitution, a variety of personal competences are required. This includes professional as well as methodological, personal and social competencies (cf. Lehmann & Nieke, 2000). The acquisition, application and personal development of the various areas of competence is a dynamic process and often exhibits fluid transitions.

In the context of prostitution, the counsellor is required to have broad-based professional competence. Legal knowledge of the content and development in the area of the Prostitution Act (Prostitutionsgesetzes) is just as necessary as, for example, in social law. Especially women entering this field of activity must be aware that the legal aspects that are needed in everyday professional life are only partially taught within the course of studies. In cases where, for example, it is a question of accompanying women in their exit from the milieu, claims have to be asserted which are often beyond the personal knowledge of the client and which can then only be asserted with the support of the counsellor. The high complexity of the legal situation in the field of prostitution means that a large number of women are not aware of their rights: "No other profession is regulated by so many different laws" (Albert & Wege, 2011, p. 10). Due to the clear observable increase of women from Eastern Europe working in prostitution, it is also necessary to be familiar with further legal foundations, such as in the area of the Aliens and Immigration Act. At the same time, these legal requirements are relevant for other areas of the women's lives, e.g. when it comes to taking up gainful employment subject to social insurance contributions or establishing their centre of life in Germany. The counsellor's specialist skills often have to be applied in the wide-ranging field of different authorities. This requires effective coordination and application of specialist knowledge in order to act in the best interests of the client.

In the direct counselling process, a broad methodological competence is required right from the start. In the sense of Galuske (2011, p. 162), this involves knowledge that provides an overview of the various methods and at the same time applies a repertoire tailored to the field of work. While in the first phase of establishing contact and access to the client, initially more social space-related methods (such as the already elaborated method of street work) are used, in the long-term counselling process, more individual case-related methods (such as client-centred counselling or case management) are found. According to Carl Rogers' counselling model, client-centred counselling is primarily about positive appreciation, authenticity and empathetic understanding on the part of the counsellor. Relevant is the personal inner attitude towards the client and not the mere application of a method. The focus is on the person concerned as a human being and not on the problem (cf. Galuske, 2011, p. 182 ff.).

Especially in the context of prostitution, a reflective and unbiased attitude on the part of the counsellor is indispensable in order to be able to establish a trusting relationship with the helper. Depending on the goals of the woman concerned, the method of case management also plays a role. This involves the coordination of offers and assistance that often exist side by side and in cooperation with several professionals within the overall process and are used and applied using resources in the individual case (cf. Stimmer, 2000, p. 49 f.) The coordination of different offers can lead to the activation of a network that provides solution-oriented offers in terms of effectiveness and efficiency (cf. Galuske, 2011, p. 199 f.). A woman who is linked to a counselling centre often has the opportunity to make use of different support services (e.g. medical, material or socio-spatial services). Often there are also further bureaucratic support measures and counselling offers at different institutions. At this point, at the latest, several professionals are involved in the process, whose work steps are ideally coordinated by a case-leading professional.

With regard to self-competence, as already described at the beginning, the personal approach to the topic of prostitution and sexuality is crucial for successful work. In counselling work it happens that intimate experiences are shared that are not compatible with one's own moral or value concepts in the area of sexuality, but which must nevertheless be absorbed. Entrenched moral concepts are a hindrance and harbour the danger of black-and-white thinking on the part of the counsellor, which is particularly counterproductive in the highly complex area of prostitution. Dealing with unforeseen situations and crises, e.g. in contact with traumatised women, is also part of the everyday life of a counselling centre and requires an extremely sensitive approach in each individual case. These areas of action in particular cannot be taught realistically in theory and in the context of studies and require counsellors to be competent in the area of crisis intervention. Due to the complex problem situations that often become apparent in the course of the counselling process, a high degree of self-organisation is required of professionals in order to steer the counselling and assistance process as "case managers". At the same time, as a specialist, it is also important to distance oneself in order to avoid absorbing the feelings and problems of the other person, which are transferred to the counsellor, and thus to maintain one's own mental health. Especially in the first years of professional life, the personal demarcation from the client and their problems is a process in which one's own professional role must first be found. Social skills also have a decisive effect on the counselling and help process. It requires a reflected self-perception and perception of others, especially in order to keep the aforementioned balance between closeness and distance in view. Due to the controversial and often critical reactions of society, the counsellor's ability to deal with conflict and criticism is also required.

Social work in the field of prostitution increasingly requires intercultural self-competence on the part of professionals in all the areas of competence described. "Professional intercultural competence in social work refers to the ability to act appropriately in professional situations that are characterised by multiculturalism" (Freise, 2007, p. 237). It can be assumed that prostitutes are primarily migrants, with most coming from the Balkan countries. For this reason, the ever-increasing need for intercultural competence on the part of professionals becomes apparent here. In the work with migrant women who work in prostitution, the importance of empathy also becomes clear, which is "characterised by a deep respect for the other. (...) The other then has the experience of really being heard" (ibid., p. 238). Behaviour, attitudes and also wishes and fears of the women can often be better classified and understood through a sensitive consideration of the cultural context. This helps to avoid mistakes in the counselling and relationship work. Information material and brochures about the services offered by the counselling centre should also be formulated in several languages and in a way that is appropriate for the target group. This increases the chances of really reaching affected women and providing access to existing services.

Although social work always starts with the individual, it requires a holistic view of the environment and the entire system of the person concerned (cf. Freise, 2007, p. 239). In counselling work with a prostitute, the living environment and the social space of the person concerned play a central role and should therefore not be underestimated. If one assumes that a not inconsiderable proportion of prostitutes have to operate in various relationships of dependency, e.g. in relation to pimps or landlords, obstacles or setbacks in the counselling work should not be attributed to a lack of willingness to cooperate, but rather to these very difficult circumstances of the living environment. Language barriers also occur more frequently in the context of counselling with migrant women and must be taken into account. Conversations conducted in the presence of interpreters are therefore not uncommon. The counsellor must consciously reflect on the fact that the presence of interpreters makes the conduct of the conversation all the more complex, since one's own stresses or subtleties of what is said can only be conveyed with difficulty via third parties. In practical work, intercultural competence is not just an approach that is simply applied in the respective field of action, but rather a skill that must be acquired and constantly refined by finding one's own professional identity.

4 Challenges in Social Work with Women

The professional, methodological, personal and social skills already described, as well as the ability to develop intercultural skills, form the basis for a successful start in the social work field of female prostitution. The practical work with the affected women themselves holds its own challenges, which are placed on the counsellor on a daily basis and again and again. Due to the manifold stigmatisations that prostitutes experience in their everyday life outside the milieu, a sensitive building of trust is required in the counselling work. Since the initial contact or access to the women is very different, as explained at the beginning, the individual situation of a woman can often only be grasped in a process and not already after the first contacts. Every woman who works in prostitution has a very different profile in terms of her age, her living and working situation, and her health, family and financial situation. For the counsellor, this means that she has to assess sensitively which information can be ascertained at which point in time and when, for example, asking questions could have a counterproductive effect on the counselling process, as it would overstep the personal boundaries of the woman concerned at that moment. Here, too, the course of the conversation should always take into account the woman's cultural background and requires a healthy self-perception and perception of others. For example, it makes a difference whether the counselling is conducted with a German prostitute or a woman with a Muslim background. Cultural, religious and family backgrounds can differ fundamentally and require an individual approach.

In contrast to counselling centres that deal with target group-oriented problems, such as in educational counselling, the counselling process with prostitutes in most cases reveals highly complex and difficult problems. For example, since there are no uniform regulations regarding occupational health and safety for prostitutes, the area of health care is also considered problematic for many women (cf. Final Report Round Table on Prostitution NRW 2014, p. 48). In order to be able to withstand the workloads, many women resort to alcohol or drugs in their everyday working lives, which can then end in addictive disorders and, for example, can significantly impair integration into a "normal" everyday life or the labour market even after leaving the milieu. A large number of women also turn to counselling centres with health complaints of a physical nature, as there is no health insurance cover and the women therefore do not have free access to the medical care system. Especially in the gynaecological field and in questions of pregnancy and contraception, there is a great need for counselling and action, which is met by counselling centres in various ways, e.g. in the form of free medical consultations or free availability of condoms. Through "education, information and prevention" (ibid., p. 53), women can be encouraged in the long term to take responsibility for their own health care.

Often, great psychological stress can also be observed, both in women who are still actively involved in prostitution and in those who have dropped out, e.g. in the form of post-traumatic stress disorder. The danger of a burn-out syndrome is also present due to a variety of factors such as irregular working hours, non-existent opportunities to withdraw or a lack of privacy (cf. Albert & Wege, 2011, p. 12). In the private sphere, too, a large number of women in prostitution lead a double life, as family members and friends often do not know the true professional activity for fear of stigmatisation and exclusion. This can lead to further psychological stress, since on the one hand there can be no exchange about the current life situation and on the other hand a non-existent life construct must always be maintained externally (cf. Girtler, 2004, p. 53 ff.).

Each of the problems described can influence the course of counselling to varying degrees, either on its own or in combination with others, and at this point again justifies a sensitive approach that must always be oriented towards the individual. For example, a client who clearly shows that she is always late for appointments may have a problem with her self-organisation. This explanation usually seems obvious. In practice, however, it became apparent in this case that the woman was unable to read the written times for the appointments and thus, due to a lack of education, could not manage to meet this requirement, although she would have been perfectly capable of doing so in her self-organisation. Social work professionals therefore need an awareness of this possibility of a wide range of problems and also an awareness of the mutual interaction of the individual problem situations. Furthermore, it should be clear that the counselling process should always be oriented towards the life and goals of the woman concerned. This requires a high level of acceptance on the part of the counsellor towards the client. For example, a woman who takes advantage of long-term counselling but wants to continue working in sex work will receive just as much acceptance as a woman who wants to leave the milieu. The counsellor's own ideas of goals and values are out of place here and hinder a successful counselling process. Offers and possibilities must therefore always be conveyed in a value-free and open manner (cf. Reichert & Rossenbach, 2013, p. 5).

The exit of a woman from the milieu is always a complex process and requires patience and transparency from both sides. In such cases, existential possibilities usually have to be secured and alternative life perspectives developed in a longer course of counselling. This can include, for example, a professional reorientation, which can then involve further intermediate steps, such as catching up on school or professional qualifications. But even after a successful exit, the counsellor's work is usually not yet complete: In the sense of Dölker and Gillich (2009, p. 118 ff.), what is needed in new phases of life and after overcoming crises is "social integration", "psychological stability", "recovery competence" and a "successful way of life". These goals and requirements often show up in the women concerned after they have left the prostitution milieu. Since an exit often also means leaving a "parallel world", integration into society is required in a variety of areas, e.g. in the world of work, the procurement of new housing and the establishment of a social network outside the milieu. In order to be able to cope with the new demands in the long term, a high degree of psychological stability is required. Only when this basis has been created can a successful life be led in the long term. In all these steps, the professionalism of the social work specialists is required and they accompany this process.

5 Challenges in the Cooperation with External Actors and Institutions

For social work professionals, dealing with external actors is also part of the everyday work of counselling. Women from the prostitution milieu repeatedly present authorities with unanswered questions, since depending on the municipal context, there is a lack of experience with this comparatively small target group. Thus, conditions, demands and possibilities often have to be renegotiated in each individual case. This can lead newcomers to the social work profession to their limits. Many authorities such as the police, youth welfare offices or other municipal institutions often have internal structures and regulations which result in their own "language" and which a social worker from an external context must understand and respect. Interpreting these "languages" and dealing with them adequately often poses great challenges for newcomers to the social work profession. If a counsellor in the field of prostitution comes into contact with such an authority for the first time without previous professional experience, it can be difficult to cope with bureaucratic problems that arise. Social workers from a counselling centre also often have to perform a balancing act between client and authority by assuming a mediating function. In doing so, it is important both to counter women's fears of authorities and institutions and to professionally confront the inhibitions and prejudices of external actors. It is at this point in particular that the counsellor needs the social and personal skills already described. Experience in leading conversations, especially in difficult contexts, as well as communication and assertiveness skills are required here. Since the demands and possibilities of these women are often in the grey area with regard to the legal basis, an increased competence in the ability to cooperate is required of the professionals. In addition, as a counsellor, one may be confronted with the different personal attitudes of external actors on the subject of prostitution, sometimes even in the presence of the women involved.

Since the topic of prostitution is strongly represented in the media, it often leads to strongly polarized opinions, both socially and politically. Prostitution can therefore be seen as a "controversial, emotionally charged and highly ambivalent field" (Howe, 2012, p. 36). As a social work professional in this field, one will therefore be constantly confronted with taking a stand and positioning oneself. Finding one's own professional role in the field of prostitution can therefore only lead to one's own professional opinion through practical experience. A counsellor entering this field of activity must also be aware of possible confrontations with clients. It can happen that men of any age contact the counselling centre, e.g. to ask for help for women who have been their customers or also when it is a matter of a man falling in love with a prostitute. At this point, it is of enormous importance to make a quick assessment of the situation in order to distance oneself professionally if necessary and to nip inappropriate inquiries, approaches or possible harassment in the bud.

6 Conclusion

Hardly any other field of social work is as multifaceted as that of female prostitution in terms of the diversity within the target group and the associated areas of law. The different forms in which and under which circumstances prostitution is practiced brings with it a wide-ranging spectrum of requirements for social work. Every social worker is confronted with this spectrum when entering this field of action professionally and should therefore be aware of her personal motivation and be able to assess her abilities and resources well. In the practice of social work, it is hardly possible to gain practical experience in the counselling context of prostitution in advance, and if it is, then this must be considered in a differentiated manner depending on the location with regard to practical work and the conceptual approach. In addition, the current situation with regard to theoretical training at universities is still unsatisfactory. Basically, there is a need for professionals in this field who have no fear of contact but, on the contrary, are open to meeting and working with a still stigmatised target group in our society. This means being open to the living environment of these women and going into their surroundings, e.g. in the form of street work. In addition to the personal attitude, which must necessarily include empathy and a high level of acceptance for those affected, it also requires a variety of competencies, both in the practical as well as in the theoretical action of social work. Without the ability to overview complex problem situations, to prioritise them and to work on them in parallel within the counselling work, one will not be able to act purposefully as a social worker in this field of action. Initial professional experience in dealing with authorities, the ability to network and experience in the field of crisis intervention should already be brought along for this area. Due to the large number of migrant women, personal intercultural skills are playing an increasingly important role. The prostitution milieu is a dynamic, highly complex living environment that is constantly changing with all its facets and all the actors involved. Social work as a professional specialist must adapt to this change and constantly develop in order to ultimately reach women in prostitution and offer goal-oriented and sustainable services.

References

Albert, M., & Wege, J. (2011). Soziale Arbeit und Prostitution. Handlungsbedarf und Entwicklungsmöglichkeiten in einem tabuisierten Berufsfeld. *Soziale Arbeit, 1*, 8–17.

Dölker, F., & Gillich, S. (2009). *Streetwork im Widerspruch. Handeln im Spannungsfeld von Kriminalisierung und Prävention* (1. Aufl.). Triga.

Freise, J. (2007). *Interkulturelle Soziale Arbeit* (2. Aufl.). Schwalbach/Ts.: Wochenschau.

Galuske, M. (2011). *Methoden der Sozialen Arbeit. Eine Einführung* (9. ergänzte Aufl.). Juventa.

Girtler, R. (2004). *Der Strich. Soziologie eines Milieus* (5. Aufl.). Lit.

Howe, C. (2012). Struktureller Wandel in der Prostitution. Zwischen Hurenbewegung und Sozialer Arbeit. *Standpunkt Sozial, 3*, 35–47.

Lehmann, G., & Nieke, W. (2000). Kompetenz-Modell. http://sinus-transfer.uni-bayreuth.de/module/modul-_10brpruefen_von_kompetenzzuwachs/kompetenzmodell.html . Zugegriffen: 20. Dez. 2014.

Ministerium für Gesundheit, Emanzipation, Pflege und Alter des Landes Nordrhein-Westfalen (MGEPA). (2014). Der Runde Tisch Prostitution Nordrhein-Westfalen. Bd. 148. http://www.mgepa.nrw.de/mediapool/pdf/emanzipation/frauen/RTP_Abschlussbericht.pdf . Zugegriffen: 8. Feb. 2015.

Reichert, S., & Rossenbach, A. (2013). Wir wollen den Frauen Unterstützung geben. *Ein Gespräch. Aus Politik und Zeitgeschichte, 63*(9), 3–8.

Stimmer, F. (2000). *Grundlagen des Methodischen Handelns in der Sozialen Arbeit.* Kohlhammer.

Julia Kempl is a social worker (B. A.) and works at the counselling centre "Amalie" – for women in prostitution in Mannheim.

Social Work with Trafficked Persons for Sexual Exploitation in Germany

Naile Tanis and Tabea Richter

1 Trafficking in Human Beings for Sexual Exploitation

In order to develop an understanding of the work of the counselling centres with trafficked persons, it is first necessary to define the term human trafficking (MH). Human trafficking is a complex problem that takes place in a national and international context. In the legal sense, human trafficking is included as a criminal offence in the Criminal Code (StGB). According to §§ 232, 233 StGB, a distinction is made between trafficking in human beings for the purpose of sexual exploitation and labour exploitation. It is punished as a basic offence with prison sentences of six months to ten years. Trafficking in human beings for the purpose of sexual exploitation occurs when persons exploit a situation of coercion in the form of the use of force, threats or deception, or the so-called foreign-specific helplessness of another person, in order to induce that person to engage in or continue prostitution or sexual acts through which he or she is exploited. The coercion may take various forms and may be exercised by psychological force or by threat thereof, as well as by extortion, unlawful withholding of documents and money earned, robbery, isolation or fraud. Exploitation of a helpless position, abuse of authority and debt bondage are also forms of coercion. Taking advantage of helplessness specific to a foreign country occurs, for example, when the person concerned is considerably limited in his or her abilities and is not familiar with the customs and legal protections of the foreign country he or she lives in, for example, by not speaking its

N. Tanis (✉) · T. Richter
Berlin, Germany
e-mail: info@kok-buero.de

M. Albert, J. Wege (eds.), *Social Work and Prostitution*,
https://doi.org/10.1007/978-3-658-37761-8_10

language (Neue Zeitschrift für Strafrecht [NStZ], 1999, p. 348). Exploitation requires that the success of the perpetrator's act of determination is precisely due to the predicament or helplessness of the trafficked person, i.e. that it is causal for the success of the offence (Fischer, 2013, para. 14). However, the term trafficking in human beings is often misunderstood, as §§ 232, 233 StGB do not concern trafficking or recruitment (Renzikowski, 2011). In addition to trafficking for sexual exploitation, there is trafficking for labour exploitation and also trafficking into marriage. Trafficked persons of MH can be German as well as foreign, as it were. The causes are manifold, but the main points to be mentioned are: growing poverty in the countries of origin, restrictive migration policies of the destination countries and an extremely low risk for the perpetrators to be held responsible for their crimes, while at the same time generating immense profits. Germany has become a country of destination, transit and origin for human trafficking.

The public often has the image that physical violence was used against the victims in order to exploit them sexually. However, this contradicts both the information provided by the specialised counselling centres and the Federal Crime Report of the Federal Criminal Police Office. According to this, in 2011, 27% of the identified trafficking victims stated that they had agreed to enter prostitution. A further 39% were deceived into engaging in prostitution and 8% of victims were professionally recruited (Bundeskriminalamt [BKA], 2011, p. 11). It is reported that while many of those trafficked are not deceived about the practice of prostitution itself, they are deceived about the overall circumstances. Trafficking in human beings is often accompanied by the deliberate creation of or exploitation of existing dependency relationships. Dependency relationships arise, for example, through "fictitious" debts for entry, which must first be worked off, or through the theft of passports, etc. In addition, there are working conditions that are not comparable to those in other countries. In addition, there are working conditions that are not comparable to the working conditions in self-determined prostitution. This is the case if, for example, only a small part of the earnings remains with the trafficked persons, they do not agree with certain sexual practices per se, or they cannot select clients in a self-determined manner (Prasad, 2008, p. 70). Only then do we speak of human trafficking for the purpose of sexual exploitation.

2 Current Political Developments in the Protection of Victims of Human Trafficking as a Framework for Social Work with Trafficked Persons

The introduction of Sections 232 ff of the Criminal Code is based on an amendment to criminal law in Germany in 2005, which in turn goes back to a Framework Decision of the Council of the European Union on combating trafficking in human beings. In the meantime, the EU Directive 2011/36/EU of 05.04.2011 ("Directive on preventing and combating trafficking in human beings, and protecting its victims, and replacing Framework Decision 2002/629/JHA") exists, which should have been implemented by the Member States by April 06, 2014. The Directive requires an integrated, holistic and human rights-based approach to combating trafficking in human beings and aims to implement better protection of victims in addition to improving prosecution and prevention of crimes. Germany has not yet implemented the EU Directive. The Directive includes as further forms of exploitation the exploitation of begging, the exploitation of criminal activities as well as organ trafficking. These forms of exploitation are not yet covered by the criminal offence of trafficking in human beings in Germany and are therefore not punishable under sections 232 ff of the Criminal Code.[1]

On December 19, 2012, the German Federal Government ratified the Council of Europe Convention on Action against Trafficking in Human Beings (ETS No. 197). The law entered into force on 01.04.2013. The Convention has dedicated its third chapter exclusively to the protection and promotion of the rights of victims. The implementation and application of the Convention is monitored, on the one hand, by a Group of Independent Experts on Combating Human Trafficking (GRETA) and, on the other hand, by a Committee gathering representatives of the States Parties. GRETA prepares reports on the basis of information it receives from the States Parties, which are made available to the public in addition to the respective State Party and the Committee of the Parties. On the basis of these reports, the Committee is empowered to make recommendations to the States. A report on Germany will also be prepared by GRETA in the foreseeable future.

The essential core of both international legal instruments is that the access of trafficked persons to counselling services must be ensured before, during and

[1] At the time of writing, a draft bill of the Federal Government is available (Drs. 17/13706), which was adopted by the Bundestag in second and third reading on June 27, 2013. The draft is currently before the Bundesrat for negotiation.

within a reasonable period of time after criminal proceedings.[2] Accordingly, Germany must take measures to ensure that trafficked persons receive support and assistance. In sentence 2 of recital number 18 of EU Directive 2011/36/EU, Member States are to provide resources for the support, care and protection of trafficked persons. In addition, EU Directive 2012/29/EU of the European Parliament and of the Council of 25 October 2012 on minimum standards on the rights, support and protection of victims of crime and replacing Framework Decision 2001/220/JHA should be consulted, which includes the right of access to victim support under Article 8 and must be implemented by Germany by November 2015.

In this article, victim support centres/counselling centres refers to the specialised counselling centres for trafficked persons (hereinafter referred to as FBS). There are approximately 40 FBS in Germany (Kavemann et al., 2012). The funding of the FBS varies greatly from state to state. In general, however, it can be said that funding is too insecure and too scarce. In order to be able to continue to support trafficked persons in line with their needs, it is essential to provide the specialised and well-networked counselling centres with sufficient financial and personnel capacities. For example, it would have to be ensured that the counsellors and language mediators are qualified for the care of traumatised victims through further training and supervision. In order to be able to cope with the constantly changing phenomenon of trafficking in human beings and the new challenges and tasks associated with it, it is also necessary to expand the existing support structure. The mandates of the specialised counselling centres should be expanded if necessary and the expansion of their tasks should be promoted accordingly.

3 Specialised Counselling Centres for Victims of Trafficking for Sexual Exploitation

Self-Image

On the initiative of women's counselling centres, migrant projects, volunteer women, counselling centres for prostitutes and institutions with a church background, specialist counselling centres for trafficked persons for sexual exploitation have been established in the Federal Republic since the end of the 1980s. Through the supraregional networking of these FBS in the KOK (Bundesweiter Koordinierungskreis gegen Menschenhandel e. V.) and the growing political im-

[2] Art. 11(1) EU Directive 2011/36/EU and, in the broadest sense, also Art. 12 of the Council of Europe Convention ETS No. 197.

portance of the problem of human trafficking, social work with trafficked persons has gradually become more professional (Rabe, 2012, Chap. I, p. 1).

The FBS organized in the KOK e. V. are non-governmental institutions under various sponsorships or from the autonomous women's movement. The social work of the FBS focuses on those affected by human trafficking and thus by a violation of human rights, especially women. In 2008, the FBSs united in the KOK agreed on four guiding principles that shape the self-image of the FBSs and thus their work: Clear positioning (advocacy and setting impulses for social and political change), partiality for those affected, professional competence, and an anti-racist and anti-sexist approach (Rabe, 2012, Chap. II, p. 1).

The FBS point out grievances in practice and use them to develop demands and political recommendations for action with the aim of influencing the social conditions that foster human trafficking and violence against women and migrants and improving the living conditions of trafficked persons: "Without the systematic linking of individual case work and political work, the work of the FBS would suffer from the fact that only 'symptoms would be alleviated', but causes would not be addressed" (Rabe, 2012, chap. II, p. 2). At the same time, it should be noted in these political activities that the demands of the FBS are not used to justify stigmatisation and disenfranchisement of certain social groups (e.g. restrictive measures in migration policy). The FBS are committed to ensuring that trafficked persons are not instrumentalised as witnesses for more effective criminal prosecution, but are strengthened and protected in their rights (Rabe, 2012, Chap. II, p. 2). This also includes that the FBS deal with the issues of discrimination, racism and sexism (self-critically!) and enforce this in concrete partisan support for trafficked persons and in political work vis-à-vis social actors (Rabe, 2012, Chapter II, p. 4).

Technical Competence

Social work with trafficked persons requires technical competence and a specialisation of counselling work in order to be able to carry out the work in its complexity in an appropriate and professional manner. The professional competencies of FBS staff include, for example, knowledge of theories and methods of social work, context-specific knowledge of human trafficking and migration, basic legal knowledge, psychosocial counselling basics, knowledge of health aspects, skills in cooperation, networking, public relations and lobbying, project management and fundraising, administration and concept development, sensitivity to discrimination and ideally multilingualism (Rabe, 2012, Chapter II, p. 3 and Chapter V, p. 17).

Partiality

In the KOK manual, partiality is understood as follows: FBS act solely in the interest and according to the needs of the trafficked persons and preserve their autonomy of action. The goal is to strengthen trafficked persons in their often disadvantaged and discriminated position and to promote their self-determination and autonomy. Critical partiality has in mind neither to victimize nor to idealize the trafficked persons (Castro Varela, 2006, p. 61; cited in Rabe, 2012, chap. II, p. 3). While the responsibility for acts of violence inflicted on the victims is clearly attributed to the perpetrators, the victims are at the same time encouraged and strengthened in taking responsibility for their own lives. In the context of the FBS, partiality also means revealing structural disadvantages and real limits. At the same time, those affected are supported in exercising their rights within the existing legal system (Rabe, 2012, Chapter II, p. 3 f.).

Since FBS work together with many cooperating institutions such as the police, the employment agency and immigration authorities in counselling and supporting those affected, it is important to communicate this partiality for the cooperation partners as well as for the affected persons themselves and to make the different roles transparent. Partiality as a guiding principle of the social work of the FBS is challenged again and again in everyday work, e.g. by the interests of the responsible bodies, the interests of social groups and the law enforcement interests of the state. The work of the FBS must therefore be constantly re-evaluated with regard to the question of how the actual interests of those affected can be taken into account in the work and how those affected can be supported in their decision-making autonomy.

Targets

In 2008, the FBSs that are members of the KOK formulated six common goals that they want to achieve through their work:

1. *Sustainability/long-term and future-oriented work*
   The aim is to sustainably improve the prospects of trafficked persons by strengthening their life management skills and asserting human rights in the individual counselling process as well as by safeguarding human rights in the political work of the FBS. The political work of the FBS aims to counter human rights violations with effective and long-term political, social and legislative

measures and to prevent human trafficking. The FBS oppose discrimination, racism, exclusion and stigmatisation of migrants (Rabe, 2012 Chapter III, p. 1, Lob-Hüdepohl, 2007, p. 134; cited in Rabe, 2012, Chapter III, p. 1). This is done, for example, through joint statements by KOK e. V. on draft legislation affecting the target group.

2. *Improving the lives of Those affected*
   This goal is pursued at the level of individual counselling with the respective person concerned as well as through the general influence of the socio-political work of the FBS. In this context, the FBS are particularly active in supporting the improvement of objective living conditions through improved legal framework conditions as well as through concrete measures, such as secure accommodation, securing livelihoods and, if necessary, residence status as well as medical care. Securing basic needs and objective living conditions can then also contribute to improving the subjectively perceived living situation (Rabe, 2012, Chapter III, p. 2). The participation of those affected in these measures is imperative, because those affected ultimately define for themselves what they perceive as an improved living situation. Improving the living situation of those affected is directly linked to the next goal of helping them to help themselves and empowerment.
3. *Help for Self-Help and Empowerment*
   The specialist counselling centres do not work for the victims, but together with them. Those affected have often lived in relationship structures characterised by dependency and exploitative care ("I'll take your passport so that you don't lose it"; "You give me the money so that I can save for us and go shopping for you"; "You have to stay in the flat because you can't manage on your own outside").

   In counselling work, it is important not to confirm these relationship structures and the self-images associated with them, but to encourage the clients to act on their own initiative and responsibility and to accompany them. This is challenged in practical work, for example, by the fact that those affected often experience that they cannot assert their interests without the support of the specialist counselling centres, e.g. vis-à-vis job centres. In addition, it is important to take into account the possible traumatisation of those affected.[3] In counselling work, care must therefore be taken to ensure that those affected are not overburdened by the demands of self-help, but that the process is supported by needs-oriented assistance (Rabe, 2012, Chapter III, p. 3).

[3] Katarzyna Zentner (2009) writes on the traumatization of trafficked persons in "Mensch im Dunkel- eine qualitative Fallstudie zu osteuropäischen Opfern von Frauenhandel" Frankfurt am Main: Lang.

4. *Safeguarding the Work of FBS*
The work of the specialised counselling centres is repeatedly characterised by the struggle for sufficient financial resources to carry out their own work. The protection and support of trafficked persons is a social and political obligation. Adequate funding and a nationwide expansion of the counselling centres are therefore urgently needed (Rabe, 2012, Chapter III, p. 3).
5. *Networks and Cooperation*
Through cooperation and networking, trafficked persons should be supported according to their needs and a broad public should be made aware of the issue of human trafficking. Furthermore, networking should help to continuously place the issue of human trafficking on the political agenda and implement a holistic, interdisciplinary and society-wide approach to human trafficking (Rabe, 2012, Chapter III, p. 4).

Interdisciplinary networking is necessary, e.g. with women's shelters, youth welfare offices, educational institutions, (victims') lawyers, medical institutions, the police and the employment agency, in order to support those affected according to their needs.

At the same time, they are often the ones who refer trafficked persons to specialised counselling centres. Therefore, continuous sensitisation of the staff is necessary. This is achieved, among other things, through specialist days and lectures at training courses for the individual professional groups and round tables. Intradisciplinary networking between the individual FBS often takes place at the level of the federal states. In North Rhine-Westphalia, for example, various RABs have joined together in a network. In Baden-Württemberg, there is a working group "Active against trafficking in women and exploitation",[4] which sends out an information mail twice a year with current developments and legal changes on the topic of human trafficking to various actors from the police, legal profession, administration, etc. to raise awareness.

At the federal level, intra-disciplinary networking takes place through KOK e. V. The work in these various networks gives the FBS the opportunity to clarify identified problems at a higher level as well as to promote general support and knowledge for practical work.
6. *Prevention*
Prevention in the work of the counselling centres includes individual and general prevention. Individual prevention means strengthening and supporting the individual trafficked person in order to prevent him/her from becoming a victim of human trafficking or other forms of violence again.

[4] http://www.aktiv-gegen-frauenhandel.de

Raising awareness and educating various actors on the subject of human trafficking also contributes to prevention. The FBS also advise politicians and administrators on the establishment of appropriate prevention measures (Rabe, 2012, Chapter II, p. 4).

If properly equipped and financed, FBS also carry out prevention programmes, e.g. in cooperation with non-governmental organisations in the main countries of origin of trafficked persons (including Germany). Examples include the projects "Open for young women",[5] "An informed person is a protected one",[6] "Lost in Cyberworld"[7] and prevention projects in schools on violence in (young) relationships and sexual assault. In this context, it is important not to view human trafficking as a stand-alone criminal phenomenon, but as a human rights violation that also takes place in the context of violence in relationships, violence in the social environment, and in the context of structural exclusion and discrimination of social groups.

Offers from Specialist Advice Centres

The services offered by the FBS are very diverse and in part very different. However, most of the FBS offer public relations, networking, lobbying and committee work as well as direct support for those affected. In the following, the focus will be on the support of the clients.

- *Psychosocial counselling and support*
  Counselling for trafficked persons is free of charge, confidential and voluntary. Counselling "focuses on the subjective reality of the client's life – her thoughts, feelings and actions – and at the same time deals with the influencing social conditions. [...] The aim of psychosocial counselling is to promote and (re)establish the affected women's competences for action, decision-making and coping, including the usable legal spaces" (Rabe, 2012, Chapter VI, p. 3 f.). Counselling is oriented towards the concerns, goals and individual support needs of those affected, their current life situation and their background experience (Rabe, 2012, Chapter VI, p. 3). At the same time, this means that during

[5] http://www.open-for-young-women.org

[6] http://www.vij-stuttgart.de/images/pdf/wer_informiert_ist.pdf

[7] http://www.lostincyberworld.eu/

the counselling interview, the counsellor must keep in mind the current legal situation of the person concerned (e.g. residence status), their state of health (e.g. traumatisation), cognitive capacities (Can the person concerned concentrate on the interview at the moment? Are there any cognitive limitations that need to be taken into account?) as well as their language skills (Is interpretation necessary? What vocabulary and background knowledge do the clients have? Do "officialese", technical terms, unknown terms and concepts need to be explained? etc.).[8]

In the following, the different phases and forms of counselling will be presented on the basis of a case study of Ms. S.

- *Background and contact*
  Ms. S. comes from a family with many children and lives in one of the new EU Member States. The parents are often absent and alcoholic, violent attacks occur again and again. Ms. S. takes care of her younger siblings a lot and leaves school early to support her family. When she can no longer stand her father's violence and the family's financial situation becomes increasingly precarious, she decides to go to Germany in search of work. She is able to stay with a family in a big city for a while and work. When conflict arises, she is kicked out of the apartment. With no place to stay, no job and no knowledge of German, she doesn't know how to help herself. An acquaintance of the family meets her and offers her a job in a brothel. Ms. S. accepts. When the acquaintance introduces her to the brothel owner, he receives several thousand euros. The brothel owner provides her with living space and informs her that she has to work off the money they paid for her. Ms. S is stripped of her ID and locked in an apartment. She is told that she will go to prison if she is picked up by the police without ID. When she does not actively approach the customers while working in the brothel she is beaten by the brothel owner. When one day she forgets to lock the door, she flees from the apartment and turns to the police. Shortly afterwards, the investigators in charge informed the victim protectors and an interpreter. The victim protectors inform her about possible support options and Ms. S. agrees to contact the FBS. The victim protectors ask the specialist counselling centre about counselling and accommodation for the victim.

In the case study described here, contact with the specialist counselling centre is established through the police. Often, however, contact is also made through other

[8] Information on the concept of "Leichte Sprache" can be found at www.leichtesprache.org

agencies such as general social counselling, social counselling in hospitals and psychiatric clinics, pregnancy counselling, women's shelters, social welfare offices or through private individuals and the affected persons' own enquiries. The prerequisite for this is raising awareness among the general public, but especially among the police, other counselling centres and offices, and also among potential victims, e.g. by providing information to people working in the sex industry about their rights and support options (e.g. in cooperation with counselling centres for sex workers).

- *Initial counselling and crisis intervention*
  The counselling centre offers Ms. S. an initial consultation with the support of an interpreter. She is introduced to the role of the counsellor, the working principles of the FBS and the counselling and accommodation services offered by the specialist counselling centre. This also includes information on how the specialist counselling centre handles her data and that she has access to her file at all times.[9] Ms. S. is also informed about her rights as a trafficked person and receives an information sheet in her language explaining the rights for victims of crime as well as information about the Victim Compensation Act and other compensation rights.[10] She can read through the information sheets again later at her leisure. Ms. S. is very sceptical at first, she asks, for example, what "women's shelter" means and "whether she will be locked up there". These fears can be allayed in the conversation. Possible support and action options are pointed out to her and it is discussed with her which steps she would like to take. It is also clarified whether she needs urgent medical support or whether any injuries still need to be documented[11] and how psychologically stable she is. After the

[9] Further information on standards on data protection and informational self-determination of trafficked persons is available on the homepage of the data protection project "datACT": http://www.datact-project.org/startseite.html

[10] As part of the EU-wide COMP.ACT project, KOK e. V. has produced an information flyer that provides information on various compensation options in Germany. The flyer can be found at the following link: http://www.kok-buero.de/kok-informiert/aktuelle-kok-publikationen/informationsmaterial.html

[11] The documentation of possible injuries is often already done by the police. If contact is made with the counselling centre without prior contact with the police, it is advisable to contact a violence outpatient clinic in the event of injuries, which documents injuries in such a way that they can later be proven in court proceedings if desired. There are various models for this in the federal states, such as the violence outpatient clinic of the Institute of Legal Medicine at the University of Heidelberg (http://www.klinikum.uni-heidelberg.de/Gewaltambulanz.130412.0.html) in Baden-Württemberg since 2012.

interview Ms. S. would like to take advantage of living in a shelter and the support services offered by the specialist counselling centre. She is placed anonymously in a shelter at the specialist counselling centre. She has her own room, which she can lock, and she can also watch TV and listen to music. She shares a bathroom and kitchen with other women. She also receives hygiene articles, an initial supply of clothes and a food parcel. She receives a new mobile phone and a new SIM card so that she cannot be found by the defendants via her old mobile phone number or the IMEI number of her old mobile phone. It will be discussed with her how to take care of her protection and anonymity while making calls and using the internet. She is also given the service mobile number of the counsellors at FBS who she can contact if she has any questions or is in crisis. An appointment is made for the next day.[12]

The example shows the possible course of an initial consultation. Building trust is central. Initial counselling often involves establishing the background and current life situation of the client, explaining the working methods and principles of the FBS, providing information and pointing out possibilities for counselling, up to and including crisis intervention. The concerns and counselling needs of the client are elicited. Depending on whether the person concerned wants further support, appointments and agreements are then made with him/her and the next steps and support measures are initiated (Rabe, 2012, Chapter VI, p. 10).

The fact that those affected get involved in counselling and initial support measures is an enormous leap of faith. It is therefore all the more important to make the working methods and principles of the FBS tangible for the clients, to show reliability and to discuss the individual steps with them.

- *Stabilisation and support in clarifying the legal situation*
  The next day, the counsellor discusses the next steps with Ms. S. with the support of an interpreter. She explains to Ms. S. once again which social benefits she can claim and what is necessary for this. Together with Ms. S. she fills out an application for unemployment benefit II[13] and has Ms. S. give her written consent that she may contact the relevant authorities in order to support Ms. S.

[12] A helpful guide is the "Leitfaden zu Gefahren im Umgang mit Kommunikationsmedien" (Guide to dangers in dealing with communication media) of the LKA Baden-Württemberg and the Ev. Gesellschaft Stuttgart (eva) and can be requested from Terre des Femmes.

[13] It is often necessary to send at least a brief notification of need for assistance to the job centre on the very first day of counselling and admission in order to ensure that all claims under social law are met.

> in claiming social benefits. Ms. S. has already made it known to the police that she would like to make further statements and would also make herself available as a witness. The police then organise a residence permit for trafficked persons in accordance with § 25 (4) a of the Residence Act (Aufenthaltsgesetz) at the responsible foreigners authority. The counsellor contacts the responsible job centre and informs a case worker about the legal regulations for trafficked persons and sends the application in order to achieve a quick clarification of costs. With the support of the police, blocking notes are created for Ms. S's files in order to ensure the greatest possible protection. In the following days, Ms. S. is also informed and supported in obtaining the necessary papers for an application (registration, pension number, job-seeker's registration, choice of health insurance, etc.). During all these visits to the authorities, special attention is paid to the protection of Ms. S.'s data, e.g. by applying for information blocks. After several weeks a clarification of costs can be achieved.

During this time, Ms. S. can take part in various leisure activities with other women, such as cooking together. She is asked what her hobbies are and what does her good in order to be able to make her appropriate offers. Over the next few weeks, Ms. S. completes several donated 1000-piece jigsaw puzzles. Putting the puzzle pieces together and finishing the puzzle does her good and helps her sort herself out.

Ms. S. regularly attends counselling sessions, which stabilise her noticeably. She is accompanied to doctors to clarify health problems. There too, attention is paid to blocking and the special protection of her data.

In the meantime, Ms. S., supported by the counsellors' networks, decides to see an experienced lawyer. She receives an initial legal consultation and comprehensive information about her rights as a victim, her right to compensation and her rights as a joint plaintiff. She authorises the lawyer to represent her as a joint plaintiff.

The counsellors of the FBS support trafficked persons in protecting their rights. This initially means providing general information about their rights and possible social law and compensation claims, support in filing applications and obtaining necessary documents and records, accompanying them to authorities and arranging for legal advice from lawyers. These often bureaucratic steps are usually very challenging and also frustrating for those affected. Again and again it may be necessary for the counsellor to explain the necessary steps and make the structures transparent. At the same time, it is important not to simply do these things for the clients, but to support them in doing this themselves as far as they are able. To feel that things are moving forward, that more and more steps are being taken, that something has already been achieved today – all this contributes to the stabilisation and empowerment of those affected. To this end, the counsellors reflect on the successes with the clients and celebrate together, for example, the social security card received, the registration that has taken place, the fact that someone at the job centre has finally declared themselves to be responsible, etc. Since an actual clarification of costs unfortunately often takes several weeks in practice, possibilities should be found as to how the persons concerned can receive help with living costs in the meantime.

Stabilising those affected and at the same time keeping an eye on security is an area of tension in the work of the FBS. For example, withdrawing into a sheltered flat, giving up looking for a job or staying away from a German course can mean the greatest possible security for those affected, but can be an obstacle to psychological stabilisation. The experience of meeting new people, going out or doing an internship contributes to the psychological stabilisation of the persons concerned, but at the same time can endanger their security. These issues are also addressed in the counselling process, encouraging clients to make responsible decisions for themselves.

- *Legal assistance, support for compensation/civil proceedings*
  In the meantime Ms. S. is questioned several times by the police. Her representative for the incidental action accompanies her. When the brothel owner is arrested and charges are brought against her, a counsellor goes with Ms. S. to a court hearing as a spectator in preparation for a trial. Ms. S. gets to know the courtroom and the seating arrangements of the individual actors. This enables her to gain some confidence in the situation, which is still unfamiliar to her. When Ms. S. is called as a witness, she is picked up by the victim protection workers and can wait for her questioning in a witness room. A staff member of the counselling centre accompanies her, and there is enough food, drink and an MP3 player with music to stabilise her. After Ms. S.'s testimony, the brothel owner receives a suspended prison sentence and has to pay a fine to Ms. S., which roughly corresponds to the amount she withheld from her in wages.

  Ms. S. does not want to initiate further civil or adhesion proceedings against the brothel owner. With the support of the counsellors and her incidental action representative, she files an application for compensation under the Victims Compensation Act.

The staff of the FBS support trafficked persons in criminal and civil proceedings and inform them about the possibilities of compensation. This includes, among other things, the organisational preparation of trials, the provision of information on the course of proceedings, the arrangement of security measures with the police, support and stabilisation on the day of the trial and professional psychosocial support during the trial.[14] In addition, they inform the victims about the possibilities of compensation, e.g. according to the Victim Compensation Act, and, in cooperation with lawyers, about civil law procedural possibilities. The court proceedings are often a very stressful phase for the victims. Insomnia, flashbacks as well as fears or panic attacks occur more frequently during this period, especially in relation to meeting the accused in the courtroom. Possible strategies for relief should be discussed with collateral counsel. The court process can be very empowering for victims when defendants receive what is perceived to be appropriate punishment based on their testimony. However, the court process can also be very disempowering if, for example, those affected are not taken seriously, there is an acquittal of the accused, etc. It is important that the results of the proceedings are communicated to the persons concerned in a comprehensible way and that they are well explained. The wishes of the victims are central to this: Would they like to testify or should the representatives of the incidental action rather work to ensure that the victims do not have to testify in person? Do they wish to make a claim for compensation on the basis of experienced health or psychological damage? The enforcement of compensation is an important political concern of the FBS. At the same time, this means that those affected also feel that they have been physically or psychologically harmed and must credibly present these harms in applications and in turn identify themselves as victims. Admitting to having been harmed means, for some victims, conceding the power to harm to the perpetrators. The FBS are therefore required to inform and advise the victims well in this regard, to respect their own decisions and to keep the options open for them to later change their mind.

- *Development of prospects and qualification in Germany*
  Ms. S. would like to learn German and look for work in Germany. The counsellors support her in applying for an integration course at the Federal Office for

[14] KOK e. V. has prepared a position paper on psychosocial support for victims of human trafficking, which is available at the following link: http://www.kok-buero.de/kok-informiert/website-news/detailansicht-website-news/artikel/kok-positionspapier-zu-psycho-sozialer--prozessbegleitung.html

Migration and Refugees. If the application is approved, the counsellors clarify with Ms. S. and the German course provider how Ms. S. can protect her data in the German course and what she can tell. Ms. S. takes part in the German course for a while, but then has difficulties keeping up with the other participants in the course. She would like to drop out of the German course and focus on finding a job. In consultation with the counsellors, she is supported in completing an internship with the cooperation partners of the counselling centre.

The counselling centre staff also support Ms. S. after the court proceedings in her arduous search for work and accommodation. Ms. S. has no training, no school-leaving qualifications and hardly any work experience and has to apply for a work permit together with potential employers because of her restricted freedom of movement. The staff of the FBS inform her about further educational paths and possibilities of integration into the labour market. However, Ms. S. does not see these opportunities as a chance for herself, but feels enormous pressure to find work as quickly as possible in order to be able to send money to her siblings and to perpetuate her stay in Germany, which is valid for six months at a time. When she finally gets a promise of a job and the employer fills out the application for a work permit with her, it takes several weeks for the work permit to be issued. The employer cannot wait that long and hires another worker in the meantime. The search for work is enormously frustrating for Ms. S.; she repeatedly falls into unstable phases in which her frustration is discharged in aggression towards her environment. The staff confront her with her behaviour and offer stabilising leisure and group activities. After some time, Ms. S. succeeds in getting a marginal job. She is additionally able to receive SGB II benefits, but is still looking for a full job. With the support of the counsellors she changes to the integration course of another school, which offers a course with a slower learning progression. Over time, Ms. S. has also built up a small social network, through which she eventually finds a room for sublet and is able to move out of the sheltered housing.

The counsellors of the FBS work together with cooperation partners in order to be able to offer a network of qualification and work opportunities for the persons concerned. This includes, for example, cooperation with providers of German courses, with vocational schools and with providers of low-threshold internships, support in the recognition of foreign vocational qualifications, in the search for work and in the preparation of CVs and applications. In doing so, the counsellors make sure that the trafficked persons are not stigmatised and that their history as a trafficked person does not become known. In the process of developing perspectives, the trafficked persons are also supported in perceiving and expanding their abilities and resources.

Trafficked persons with foreign citizenship (EU citizens and third-country nationals) receive a residence title in Germany in accordance with Section 25 (4a) of the Residence Act (Aufenthaltsgesetz – AufenthG) or are considered to be entitled to freedom of movement. However, their legal residence and thus also their alimentation through social benefits are primarily linked to their capacity as witnesses in criminal proceedings concerning trafficking in human beings.[15] In accordance with Section 26, paragraph I, sentence 3 of the Residence Act, residence under Section 25, paragraph 4a of the Residence Act (Aufenthaltsgesetz) is only granted for a period of six months and can be extended. Under certain circumstances, trafficked persons may also be granted a residence title under Section 25(4b) of the Residence Act (Aufenthaltsgesetz) in order to sue for compensation if this would be considerably more difficult from abroad.[16] Even in the case of EU citizens, alimentation is generally only granted for as long as they are needed as witnesses in criminal proceedings. During this time, the persons concerned can apply for a work permit without a priority check (§ 6a BeschverfV). As of July 01, 2013, the regulations on the Employment Procedure Ordinance were transferred to a new version of the Employment Ordinance. According to the new ordinance, the employment of foreigners who hold a residence permit for reasons of international law, humanitarian or political reasons (§§ 22 ff AufenthG) no longer requires the consent of the Federal Employment Agency (§ 31 BeschV). However, it should still be noted that access to the labour market is more difficult for the persons concerned due to the short duration of the residence permits issued. Many of those affected are also unable to look for a job at an early stage due to trauma or for security reasons. For trafficked persons who want to stay in Germany, there is often enormous pressure to find a job as quickly as possible in order to be able to stay in Germany, even after legal proceedings, not only to look for a job, but if possible already as an employee and to receive social security. If, for example, the EU citizens concerned have not found a job until after the conclusion of court proceedings, they are partially excluded from receiving unemployment benefit II if they continue to stay in Germany only to look for a job and are therefore no longer covered by health insurance.

These political conditions put pressure on the people concerned as well as on the counsellors. At the same time, the counsellors sensitise the people concerned

[15] This is also explained in more detail in the technical notes of the Federal Employment Agency on § 7 SGB II in the version of 20.11.2011.

[16] However, there is criticism here that Section 25(4b) of the Residence Act (Aufenthaltsgesetz) only provides for the possibility of extending the residence title to enforce wage claims solely for illegally employed third-country nationals and not for trafficked persons for sexual exploitation or labour exploitation, thus violating Article 3 of the Basic Law (Rabe, 2011).

not to enter into exploitative or precarious working conditions, even if they are looking for work.

- *Development of perspectives and support for the return to the country of origin (illustrated by the case study of Mrs. K.)*
  Ms. K. lives in her country of origin with her young daughter, whom she leaves with her parents when she receives a job offer as a service worker in Germany. In Germany her identity card is taken from her and she realises that the supposed job offer means that she is to work as a prostitute in a brothel. She quickly resists, is able to escape with the help of a taxi driver and is taken by him to the police, who contact the specialist counselling centre. Ms. K. wants to return home to her daughter as soon as possible. The counselling centre supports her in claiming a return grant, through which her return journey can be financed and booked. In addition, the counsellors give her addresses of non-governmental counselling centres in her country of origin so that she can contact her local colleagues if she has any difficulties or needs counselling. She can contact a counselling centre near her home town directly and arrange with them to be picked up from the airport and receive counselling on the spot in order to clarify whether and what form of further support she needs.
  As she no longer has an identity document with her, arrangements are also made with the consulate so that Ms. K. can obtain a travel title for her return as quickly as possible. The staff accompany her to the consulate to collect her travel title and then to the airport. Ms. K. has never flown before and needs support to find her way around the airport in this stressful situation. After the return journey, the counsellors contact Ms. K. again with the help of an interpreter to make sure that she has arrived safely.

For many persons of foreign nationality, the question of returning to their country of origin arises. The fact that residence is limited under Section 25 (4a) of the Residence Act (Aufenthaltsgesetz) and is not extended after court proceedings or the discontinuation of preliminary proceedings is problematic. This inevitably raises the question of a possible return for many of those affected. The FBS support those affected, if they so wish, in their return or also in achieving a longer-term stay in Germany. Some of those affected would like to return as quickly as possible, so that counselling centres primarily provide support here in organising the financing and the course of the return journey and, if desired, in establishing contact with a specialist counselling centre in the countries of origin of those affected. The KOK's extensive database of contact points is available to the FBS in their search for a suitable counselling centre.[17]

During the counselling, future prospects and alternatives in the country of origin are discussed and questions regarding the safety of the persons concerned on site are clarified. For example, it is discussed which security measures can be taken if necessary. If the persons concerned do not have any identity documents with them, it is necessary to contact the relevant consulates. Sometimes it is very difficult to get a new travel document. The counsellors support the persons concerned in ensuring that they are not discriminated against in the application process and that they are able to present the necessary documents, e.g. faxes with certified birth certificates, etc.

If there is more time to plan the departure, the counselling centres also support longer-term planning for the future. In further counselling sessions, for example, the qualifications and skills of the persons concerned are determined and what possible and realistic future prospects are open to them. If the persons concerned need further medical or therapeutic care, this should be ensured before their return. Here, too, the FBS clarify all steps with the persons concerned, obtain declarations of consent for all information to be passed on and pay particular attention to the protection of data and informational self-determination of the persons concerned when supporting their return. If the victims wish to continue to make themselves available as witnesses in criminal proceedings or to assert claims for compensation in Germany, the specialist counselling centres maintain contact with the victims and ensure that the necessary applications are made or powers of attorney are signed before departure.[18]

The specialised counselling centres support trafficked persons in their return if they so wish. They take a stand against deportations and oppose trafficked persons being taken into custody due to missing documents.

---

[17] The contact point database contains a list of counselling centres abroad as well as experience reports on cooperation with these counselling centres. The database is only accessible to member organisations of the KOK e. V. and can be accessed by them on the KOK e. V. website.

[18] In order to strengthen data protection and the right to informational self-determination of those affected, KOK e. V. has launched the data protection project "datACT" in cooperation with La Strada International. Further information on the project can be found at http://www.datact-project.org/startseite.html

4 Outlook

Human trafficking is a human rights violation. Social work with trafficked persons for the purpose of sexual exploitation stands for comprehensive counselling and support of trafficked persons on the basis of a comprehensive human rights approach. The goal is to restore or maintain the physical and psychological integrity of trafficked persons, regardless of their willingness to testify as witnesses. In order to create the necessary framework conditions for this, it is essential that the far-reaching requirements of the EU Directive on Preventing and Combating Trafficking in Human Beings and Protecting its Victims and the Council of Europe Convention on Action against Trafficking in Human Beings be implemented in Germany.

References

Bundeskriminalamt (BKA). (2011). *Bundeslagebild Menschenhandel 2011*. Wiesbaden.

Fischer, T. (2013). *Strafgesetzbuch und Nebengesetze* (60. Aufl.). Beck.

Kavemann, B., Helfferich, C., & Rixen, S. (2012). *Bestandsaufnahme zur Situation der Frauenhäuser, Fachberatungsstellen und anderer Unterstützungsangebote für gewaltbetroffene Frauen und deren Kinder*. Gutachten im Auftrag des Bundesministeriums für Familie, Senioren, Frauen und Jugend. Berlin.

Neue Zeitschrift für Strafrecht (NStZ). (1999). *BGH, Beschluss vom 03.03.1999*. Jhg. 19 (7). München: Beck.

Prasad, N. (2008). Menschenhandel in die sexuelle Ausbeutung. In Bundesweiter Koordinierungskreis gegen Frauenhandel und Gewalt an Frauen im Migrationsprozess e. V. [KOK] (Hrsg.), *Frauenhandel(n) in Deutschland* (S. 67–76). Berlin.

Rabe, H. (2011). Stellungnahme zum Gesetzesentwurf der Bundesregierung zur Umsetzung aufenthaltsrechtlicher Richtlinien der Europäischen Union und zur Anpassung nationaler Rechtsvorschriften an den EU-Visakodex. Deutsches Institut für Menschenrechte. http://www.institut-fuer-menschenrechte.de/uploads/tx_commerce/stellungnahme_zum_gesetzesentwurf_der_bundesregierung_zur_umsetzung_aufenthaltsrechtlicher_richtlinien_der_europaeischen_union_und_zur_anpassung_nat_rechtsvorschriften_an_den_eu_visakodex_01.pdf. Zugegriffen 19. Juli 2013.

Rabe, H. (2012). Sozialpädagogische Praxis – Teil 1. In Bundesweiter Koordinierungskreis gegen Frauenhandel und Gewalt an Frauen im Migrationsprozess e. V. [KOK] (Hrsg.), *Handbuch zur Aus- und Fortbildung und Qualitätssicherung für Fachberatungsstellen für Betroffene von Frauen–/Menschenhandel*. Berlin.

Renzikowski, J. (2011). Strafverfahren zu Menschenhandel zur Arbeitsausbeutung seit Einführung des § 233 StGB. In Bundesministerium für Arbeit und Soziales [BMAS] (Hrsg.), *Entwicklung tragfähiger Unterstützungsstrukturen für die Betroffenen von Menschenhandel zur Arbeitsausbeutung*. Studie koordiniert durch den KOK e. V., (S. 29–36). Berlin.

Zentner, K. (2009). *Mensch im Dunkel – eine qualitative Fallstudie zu osteuropäischen Opfern von Frauenhandel*. Lang.

Naile Tanis is a lawyer. After working as a freelance lawyer, she took on the position of managing director for the KOK (Bundesweiter Koordinierungskreis gegen Menschenhandel e.V.) in 2004. During this time, she has accompanied various projects of the KOK e. V. and written publications and statements for the KOK e.V.. The KOK e.V. is an alliance of organisations that campaigns against all forms of human trafficking and exploitation as well as violence in the migration process. It is the only coordinating body with this focus in Germany and Europe.

Tabea Richter has a degree in social pedagogy and was a member of the board of KOK (Bundesweiter Koordinierungskreis gegen Menschenhandel e. V.) from mid-2012 to mid-2014, which advocates for the rights of trafficked persons. Tabea Richter was an employee at the Mitternachtsmission Heilbronn, the specialist counselling centre for trafficked persons.

(Forced) Prostitution: Between Voluntariness and External Determination. Insights from the Perspective of Practical Social Work in a Specialised Counselling Centre Against Human Trafficking

Katharina Kähler

1 Introduction

Prostitution and trafficking in human beings for the purpose of sexual exploitation are often directly linked in public perception and especially in media portrayals. In many cases, no differentiation is made as to whether the prostitution is voluntary or forced, or whether there is possibly an intermediate or alternating form with elements of both. Furthermore, a sweeping and lurid image of women affected by exploitation is often created, which contains a strong deficit orientation and denies those affected their independence as persons. This is certainly also due to the fact that prostitution is still not widely accepted. The evaluation of prostitution as a socially unworthy activity is still stuck in many people's heads, and a portrayal of prostitution that only takes place under duress possibly makes it easier for many people to come to terms with this issue.

K. Kähler (✉)
Berlin, Germany
e-mail: kaehler@imhb.de

M. Albert, J. Wege (eds.), *Social Work and Prostitution*,
https://doi.org/10.1007/978-3-658-37761-8_11

In this respect, there are not always clear boundaries between voluntariness and involuntariness that make a clear classification possible. This applies both to outsiders[1] and to the persons involved in prostitution themselves. It is true that there are clear legal guidelines that define trafficking in human beings and other related criminal offences such as the exploitation of prostitutes as a clear injustice,[2] and the Prostitution Act is just such a clearly formulated commitment to the recognition of prostitution under the rule of law.[3] From this comes the renunciation of former immorality, even if there is still a lot of catching up to do in implementation in order to bring about an actual improvement in working conditions in prostitution (KOK, 2008). Between the legal norms and the moral evaluation, however, there is a large grey area of different lifeworlds, within which many women working in prostitution move between voluntariness and involuntariness.

This essay is intended to provide insights from practical social work, which deals with this antagonism of voluntariness and coercion within prostitution activities, and would like to show how complex the individual life situations of those affected are from the point of view of a professional counselling centre. After a brief excursus into the interdisciplinary discourse on voluntariness and free will, the resulting dilemmas of the counsellors in dealing with life situations that cannot always be clearly assigned will be addressed, as will the factors that lead to the decision-making paradigms of women in prostitution. To this end, various influencing elements such as role models, life and migration backgrounds will be examined by way of example, but above all the professional view of a specialist counselling centre will be presented and possible approaches to dealing with the different facets of voluntariness and involuntariness in social work with (forced) prostitutes will be shown. Case studies show how heterogeneous the manifestations of trafficking in human beings can be and what practical approaches[4] there are to support them. The following section will then examine how counsellors in this field of counselling develop a professional framework for their work and how they can

---

[1] This also refers to institutions that have direct contact with women in prostitution. Although they often have a high level of professional knowledge about prostitution and human trafficking, they are not always familiar with the specific complexity of the individual life situation of each person, or often only after long counselling processes.

[2] Among others trafficking in human beings §§ 232, 233 Criminal Code (StGB), promotion of trafficking in human beings § 233a StGB, exploitation of prostitutes § 180a StGB, pimping § 181 StGB.

[3] Prostitutionsgesetz § 1.

[4] Nothing else is the task of a specialist counselling centre.

deal with the different ways in which trafficked persons view voluntariness and coercion in their professional work.

2 A Brief Interdisciplinary Excursus on the Concept of Voluntariness

The ideological discourse on the concept of voluntariness and free will in the humanities encompasses philosophical, psychological, legal and sociological dimensions, among others. There is no universally valid definition of this. However, there are various derivations, some of which are briefly outlined below and used to provide a theoretical underpinning for explanatory approaches and motivations for action in relation to the topic of prostitution and forced prostitution.

Philosophical Approach to the Concept of Voluntariness

In the philosophical consideration by Schoppenhauer, free will is defined as "that which is in no way necessary, i.e. not dependent on any reason", "without a preceding cause", "without necessity", "not determined by reasons" (Schoppenhauer, 1977), but manifests itself in humans depending on causality and motives and is coupled to thought and action (Schoppenhauer, 1977). Will and action become explicable by the character of the agent and his situation at the moment of a decision, which is determined by this (Schmidt-Salomon, 1995). Schoppenhauer assumed that the free will of man can never be completely independent of the acting person and is determined by him. He is considered the founder of determinism,[5] which implies a direct connection between a person's motivation to act and his situation and environment. Contrary to the socialisation-theoretical approach presented below, however, he assumed that the character that is causal for the acting person and his decisions is innate and thus static and does not change in the course of a person's socialisation.

[5] From Latin determinare, "to delimit", "to determine".

Socialisation-Theoretical Approach to the Concept of Voluntariness

Character development, on the other hand, is understood as dynamic in socialisation theory. Thus Hurrelmann, as one of its representatives, understands and regards socialisation as the "process of the emergence and development of the human personality as a function of and in confrontation with the social and the material-material conditions of life...which exist at a particular point in time in the historical development of a society" (Hurrelmann, 1989). He sees human beings as part of a chaotic-deterministic overall system in which they act in a self-organizing way, whereby self-organization is not interpreted as an expression of inner freedom but as the result of the history of the individual's experienced structural couplings. Individually acquired structural couplings, which lead to evaluations and motivations for action, condition the individuality of man. His decision patterns are not arbitrarily constructed, but biographically acquired (Schmidt-Salomon, 1995). Accordingly, decisions are made in the context of previous life circumstances and the resulting, dynamically changing personal views and norm and value orientations. External influences can certainly lead to a motivation for action being taken by the protagonists as their own decision, but under the pressure of existing necessities that they themselves do not specify. This assumption will be made in the following for the theoretical underpinning of the consideration of voluntariness and coercion in prostitution.

Legal Approach to the Concept of Voluntariness

The legal definition must also be included in the further consideration, since the criminal offence of human trafficking for the purpose of sexual exploitation is an essential aspect of this work, which is clearly based on a coercive situation setting that has relevance under criminal law. This makes it necessary to take a look at the legal classification. For counsellors in social work with prostitutes and possible victims of human trafficking, knowledge of the legal classification in the area of criminal offences to the detriment of prostitutes is indispensable. The basis for this is the Basic Law, which in its first article on inviolable human dignity is based on the idea of the human being as a being "who is designed to determine and develop himself in freedom".[6] This freedom of will in the sense of self-determination can

---

[6] BVerfG, judgment of 15.02.2006, Ref.: 1 BvR 357/05.

only be temporarily or permanently impaired by pathological disorders, i.e. for reasons that lie within the corresponding person himself; any other external determination by others is subject to various criminal offences. The possibility of making one's own decisions and thus acting in a self-determined manner is enshrined as a fundamental constitutional good and is thus regarded as the basis of our society. The Criminal Code is also concerned with free will, it lays down the conditions and legal consequences of criminal offences and sets a framework as to whether and to what extent executors are responsible for their actions. Prior to this, however, there is an evaluative attribution. The Criminal Code defines which acts are understood as criminal offences. These are first considered, independently of the acting person, as behaviour not conforming to the norm to the detriment of others, and it is determined that[7] this deviant behaviour is undesirable and punishable. Only after this does the reference to the executing person and the legal assessment of his or her responsibility follow. Relevant for this work is primarily the criminal offence under § 232 StGB, this falls in the Criminal Code under the 18th section, to which crimes against personal freedom are assigned. "This refers to all acts by which people of any sex or age are brought into an economic exploitative relationship in which their right to self-determination is violated and they are forced into prostitution or other sexual services through physical and psychological violence, threats, deception, blackmail, exploitation of a predicament or exploitation of their helplessness" (KOK, 2012, Chapter 1, p. 2; Kalthegener, 2008, p. 76 ff.).

The legal consideration of voluntariness and personal freedom is relevant to this work insofar as laws, as a social norm structure, have a considerable influence on the assessment of a life situation in the question of actual voluntariness.

Relevance of Discipline-Specific Approaches to Social Work

In contrast to the preceding causal approach to free will from philosophical and socialisation-theoretical perspectives, which analytically aim at a description and classification for human decision-making motivations, the legal approach includes another dimension. It defines a social norm by establishing free will as indispensable. Furthermore, it provides a sanctioning framework for society by which it determines how to respond to deviation from the norm (Spittler, 1967). In social work, both play a crucial role in counselling contexts with people who may be trafficked. On the one hand, it is necessary to be able to theoretically explain the intensions of action in order to understand the specific life situations of those being counselled

[7] Civil Code § 104.

and to be able to deal with them appropriately in the counselling process by the counsellors; on the other hand, the legal framework plays a decisive role in explaining choices and their evaluation by the counsellors. It therefore seems sensible to at least touch on the context of voluntariness in different disciplines in order to be able to base the following on a certain theoretical derivation and not just rely on practical experience.

3 The BBMeZ (Counselling Centre for Trafficked Persons and Forced Prostitution) as an Executive Body of Practical Social Work with Trafficked Persons

With regard to the action-oriented perspective of social work on the various manifestations of voluntariness and coercion in prostitution in this essay, it seems useful to provide an insight into the basic character of the working methods and tasks of practical social work at BBMeZ.

Since 2002, BBMeZ has been advising, accompanying and supporting victims of human trafficking for the purpose of sexual exploitation in Bremen as an independent specialist counselling centre. Initially planned as a temporary project at the request of a political party to the Bremen Senate, it soon became apparent that the need for counselling was considerably higher than initially assumed and taken into account by the funding. In the past years, an average of 30–40 women were counselled annually[8]; accordingly, the employees of the counselling centre were able to gain deeper insights and experience in working with trafficked persons, develop comprehensive counselling competencies and get a picture of the prostitution scene in Bremen. This should be reflected in the following discussion.

The self-image of the work includes first and foremost the acceptance of prostitution activity, provided that the decision to do so was made by the respective person of their own free will. In addition, the willingness to fight human trafficking as a massive violation of human rights, national and international law, respect for different religions and faith communities, lifestyles and societies of origin, as well as the recognition of the right to make one's own decisions in the persons advised and a resource-oriented working attitude are indispensable. The counselling centre is run by a Christian non-statutory welfare organisation, but the counselling centre does not pursue any missionary interests and sees its self-obligation in providing

[8] Annual reports of the BBMeZ counselling centre in the Verein für Innere Mission in Bremen for the years 2008–2012.

professional counselling for trafficked persons that is non-denominational and oriented towards the interests of the persons concerned. Women (and in individual cases also men) who have actually or possibly become victims of human trafficking find a socio-pedagogical counselling service here. They can use this to reflect on their own life situation, to inform themselves and to weigh up their options for action together with the counsellors. The counselling centre also offers decision-making aids, but not in the sense of external determination; instead, it helps those affected to sort through, to sound out and to implement their decisions. There is cooperation with the criminal prosecution authorities only if it is desired on the part of the persons to be counselled. Rather, the approach is to give those being counselled an overview of what legal requirements and legal claims exist and to give them the opportunity to classify their situation in life and to decide how they want to proceed. In detail, potential trafficked persons can find the following services at BBMeZ:

- First aid
- Organization of safe accommodation
- Assistance with medical care
- Administrative support and accompaniment to visits to the authorities
- Clarification of the possibilities under residence and social law
- Information about legal aid and legal options/referral to lawyers
- Psychosocial counselling and stabilisation
- Support in coping with everyday life
- Preparation and follow-up as well as accompaniment to court proceedings
- Support in the development of new life perspectives
- Support during pregnancy and maternity
- Help with language, school and vocational qualifications
- Referral to complementary counselling services
- Organisation of the departure to the home country, mediation to local counselling centres
- Financial and material support in individual cases

Within this counselling offer, those receiving counselling experience different possibilities of assistance. They decide for themselves which assistance is appropriate for their situation and which offers they accept. An important feature of the counselling centre's work is its willingness to look for unconventional solutions if necessary. Further tasks of BBMeZ lie in the area of public relations and lobbying for those affected and against human trafficking. The counselling centre sees itself as a socio-political demand maker whose task is to draw attention to the problem of human trafficking as a violation of human rights and a massive infringement of self-determination. Prostitution is not regarded as a social grievance, but human trafficking as one of its manifestations is, of course.

4 Case Studies

The following are case studies from practical social work at the BBMeZ counselling centre. They describe different life situations of affected persons who have been supported in the counselling centre in recent years. There is no claim to make an empirical typification and to show the complete range of counselling social work in the field of trafficked persons, rather the intention is to show how complex and individually different the living conditions of those who appear in the counselling contexts can be.

Case Study 1: Ms J., Born 1990, Nationality Germany

Ms. J. is 23 years old, a German citizen and comes from a small town in northern Germany. She has a secondary school diploma and lived in her parents' household until she was 17. At the age of 16 she met her boyfriend V., 12 years older than her, in a discotheque. At first she had a relationship with him, which was strongly marked by emotional dependence on her part. Her relationship with her parents had been marked by conflicts and a lack of understanding for each other for years. She left her parents' home and moved in with her boyfriend. When she turned 18, he demanded that she contribute financially to their life together. He persuaded her to work as a barmaid in a club owned by friends of his. Initially Ms. J worked as a barmaid, she refused offers from bar patrons to provide prostitution services in return for payment. She reported that at that time she had received support from her boyfriend V. and the bouncers who were friends of his to refuse such offers. After about six months, V. told her that he was heavily in debt and needed her help. He suggested that she make herself available for a certain period of time for requests at the club. She could choose the customers and also what she offered. Besides, this was only for a short time until the debts were paid off. At the same time he proposed marriage, which she accepted. Mrs. J decided, despite reservations, to at least give it a try, he had promised her that they wanted to build a future together and later, when there was enough money, to buy a house. She began to prostitute herself in the club and in apartments of acquaintances of her boyfriend, her boyfriend took care of appropriate contacts. Arguments ensued when she did not earn enough from his point of view. He began to beat her and put her under pressure, threatened her and put her under considerable psychological pressure. She saw no possibility for herself to resist him. She hardly had any trusting contacts with others, sometimes superficial contacts developed with other women in the apartments or clubs. V. made sure that she did not work anywhere for more than a week. In addition, Ms. J had broken off her own friendly contacts from the past for some time. He organised where she had to work the next week, took her to the addresses, sometimes friends of his also drove. She had to hand over most of the money, but was allowed to keep a small part for food and clothing. She reported that she had thought about fleeing, but did not dare draw anyone's attention to her situation. The relationship with V. had developed into one of pimping.

She established a closer contact with V.'s friend M. after some time, talked to him about V. when he drove her to new clubs. M. showed interest in her. One day V. informed her that he had sold her to M., who had offered him a certain amount. At that time she had already been engaged in prostitution for 1.5 years, of which initially about half a year under self-determined conditions. At that time she assumed that she was working on a common future with V. After that, she found herself under pressure for about a year and without any possibility to escape from V.. Her health and particularly her mental health were poor at this time, there was a lot of arguing with V., he demanded that she work more, she had several mental breakdowns. She suspects that V. sold her because she was no longer able to work regularly and could not meet his demands. He probably saw this as the greatest profit still to be made. She was glad when M. told her that he had "ransomed" her, but that she would have to help him work off the money he had to raise for this. This time, however, she was allowed to decide for herself under what conditions she worked, he also earned money with his job as a bouncer and security guard. Through his connections in the milieu he could protect her from V. and offer her much better working conditions, the earnings were to be shared by both of them. Ms. J. also worked for M. for some time, but only in an apartment and a permanent club. She was able to keep half of her earnings for herself, lived with M in a shared flat and later reported in counselling that she had come to terms with her situation at this point. What was demanded of her seemed bearable in relation to her previous experiences with V. She had contacts at the club with other women, worked from evening to morning, spent her days in the shared flat and saw giving away half of her money as a contribution to the shared life. She suffered from depression but managed to maintain a certain level of work ability and to lead what she saw as a fairly regular life with M. She was a woman who had a lot of experience of the red-light district. Her centre of life was the red-light milieu, which was closely linked to her relationship with M. She drew self-confidence from complicity. She drew self-confidence from compliments from clients and sometimes from M. telling her she was doing her job very well. The thought of quitting came up from time to time, especially when she was unwell and struggling with depressive episodes. She drank a lot of alcohol during these episodes to help her cope, but saw no way out and felt unable to act to get help. She shared that she had been afraid of not being able to make it on her own and in a middle-class life beyond the red-light milieu, or of not being able to build a new perspective on life.

Even if she was not currently forced by M. to continue her activity through the use of violence, there were always subtle threats that kept her from leaving. She only received support when she became pregnant by a john and wanted to have an abortion. The attending doctors at the facility where the procedure was to take place, and where M. had referred her for this, became suspicious when they gained an impression of Ms. J.'s poor mental state during the preliminary discussion for the planned procedure. Although Ms. J. was not yet able to disclose her situation there, she accepted the offer to call in the crisis intervention service of the social psychiatric service, as she was acutely suicidal at that time and recognised for herself the chance of help. The specialist service was called in and Ms. J was admitted to a psychiatric hospital. There the police were called in. Ms. J did not want to make a statement to an investigating authority, she was also psychologically unable to come to terms with what she had experienced at the time. She denied all inquiries towards forced prostitution to the police, but agreed to voluntarily stay in inpatient treatment for some time. She was transferred to a psychiatric clinic for further stabilisation. There she learned through social workers about the offer of a counselling centre for trafficked persons. She agreed to be contacted by the supervising social service and thus came into a counselling setting within the BBMeZ counselling centre, through which she was accompanied for some time. This was followed by psychiatric stabilisation with subsequent transition to a therapeutic residential group. However, Ms. J. aborted this support measure surprisingly for all parties involved. She suddenly disappeared, after being considered quite stable in the eyes of all involved, and was no longer contactable. She had left her belongings behind. Approximately six months later, the counselling centre received an email from Ms. J thanking them for the support offered and saying that she had met M again and had decided to return to him. He had promised her that she would no longer have to prostitute herself for him. There has been no contact with Ms. J since the last email, it is unclear whether she is actually no longer involved in prostitution or if so, under what conditions and with what possibilities of self-determination and influence over her own life.

Case Study 2: Ms K., Born 1995, Nationality Nigeria

Ms. K. is 19 years old and comes from Nigeria. She grew up with her family (grandmother, grandfather, mother and five siblings) in a rural area. Her father died five years ago and since then the family has been almost destitute. Her grandmother and mother occasionally work in fields for farmers in the area; Ms. K. also worked there when she was 13 years old. Until then she had attended school. Through neighbours and acquaintances she heard that one could earn money as a market woman in Lagos and that there were good contacts to Europe. A neighbour reported that he had heard that one could possibly travel to Europe to live and work there. For Mrs. K., travelling to Europe to find a job to support the family seemed to be the only option. Her older siblings were already no longer living with the family, her older brother had married and moved with his wife to their home village and was living there with her and her family, her older sister had also married and left the family. The nuclear family now consisted of Ms. K., her grandparents and mother, and two younger siblings; they lived in a slum in a hut in very poor condition and supported the family income through odd jobs. When Ms. K. told her relatives about her thoughts of trying to look for a job in Lagos, they encouraged her, especially since Ms. K.'s aunt lived in Lagos and Ms. K. could initially find accommodation there. Ms. K moved in with her aunt in the city and looked for a job. She actually found the opportunity to sort fruit and vegetables at a market and received a small income for this. With this she supported her family at home. With her aunt and her family she regularly attended church services and religiously oriented events of a church congregation, there she got to know I.. She had already learned about him that he lived in Europe for some time each year and pursued business there. What exactly she didn't know, it seemed to have something to do with cars. She sought contact with him, talked to him, and began to develop a trusting relationship with him. He told her that he was looking for young women to work as nannies in Europe. He knew people who had contacts to make appropriate placements, and the journey could also be organised. Ms. K. was happy about the offer and was not deterred by the fact that she would have to pay a high amount of money for the travel costs and the placement. She was relieved to be offered that she could repay these costs later from her earnings. She would earn well, within a year she would have paid it off and earn enough to support her family. Ms. K. agreed and she was introduced to several men who would take care of her papers and sort out the rest. A date for travel was set. Ms. K. met M., a woman who was announced to her as a contact person for the trip and as a supporter in Europe. She started her journey with her and then learned that the plan was to take her to Holland via Italy.

First Ms. K travelled by plane to Rome, from there they continued by car. Her companion M. was with her the whole time, only the drivers changed several times. They reached their destination, which turned out to be a large German city. M. told her there had been a change of plans, she was now in Germany and the promised job as a nanny was no longer. Instead, a man would be coming by later who might take her in, she was expected to be nice to him, maybe there would be a chance to work for him. Mrs. K. was intimidated when she was told that if she was not kind to him, something would be done to her family back home. She already suspected that it would be prostitution as she had been given appropriate clothing to wear. She was desperate, felt alone, did not know her way around, her identity papers were in the hands of M. and she had no money, nor was she oriented in any way. When they arrived, she had been able to make out in the dark that she was in a large house, she had been assigned a room and had gone to sleep first. The announced man came, spoke to M. and the other men, went to her room and informed her that he would take her. He spoke poor English, so M. translated for her and again threatened her that her family would suffer terribly if she did not do as he told her. It was agreed with the man called B. that he would be allowed to take her away, Ms. K. does not know whether and how much money was paid on this. She was taken to him by M. in the car at night and there she was locked in his flat. Using violence and threats against her family and her life she was forced to prostitute herself, men came every day with whom she had to sleep. In between she had to run the household, clean, wash and also be of service to B. sexually. Pornographic photos were taken of her. Again and again B. told her that she was in Germany illegally and should not be picked up by the police under any circumstances, therefore she could not go outside. During the day she was locked in the flat and B. was rarely at home. She was very unwell, desperate, very afraid and felt helpless.

After about three months Ms. K discovered that a window was not locked. She opened it and fled. She wandered through the city and was taken to the train station by a bus driver. There she met compatriots who gave her the address of an initial reception centre for refugees and explained how to get there. She somehow managed to find her way there and auditioned. She was taken in by the home's management and questioned about her situation. After she reported what had happened, the BBMeZ counselling service was notified and they went to see her and explained appropriate options. She was taken to secure accommodation, given support to clarify her situation and decided to take advantage of her statutory cooling-off period, within which she must make a decision whether to make a complaint to the criminal investigation authorities. In the end, she decided to make a report to the police, but the information she was able to provide did not lead to the perpetrators being convicted. The investigation was dropped without result and Ms. K. lost her residence status as a witness in criminal proceedings. Today she has a toleration and lives in insecure circumstances. She does not want to return to her home country under any circumstances for fear of the perpetrators. She has sporadic contact with the counselling centre when she is in great need.

Case Study 3: Ms I., Born in 1987, Nationality Bulgaria

Ms. I. is 26 years old and grew up in a suburb of a medium-sized Bulgarian city. She has two children (aged seven and eleven) who live with her parents in Bulgaria while she earns money abroad to support the children. The family is poor, they live in cramped and run-down housing conditions and can just afford the monthly rent or the bare necessities. There is no health insurance, their father is ill and one of the children is blind. The children's father is a great cousin of Ms. I., she does not know where he is staying. Ms. I. has already worked in prostitution for several years in Bulgaria and the Czech Republic to support the family financially. Her two sisters also occasionally work abroad in prostitution and send smaller amounts of money home, one sister is currently working in Spain. One day Ms. I. got an offer from her neighbour, who was also her landlord, to go with her to Germany and work in prostitution there, she would only have to pay travel expenses for the trip and a small rent for the room there. She could travel home at any time and it would not be difficult to earn a lot of money within a short period of time. This is much more than she currently has available. Mrs. I agreed. Her last stay in the Czech Republic a few months ago was not very successful financially and her reserves from that time had been used up.

She was brought to Germany in a minibus by acquaintances of her neighbour, to an apartment where three other women were already living and working. They explained to her what the working conditions were like, what she had to offer and at what prices. She also learned how much the weekly rent for the apartment was and what formalities still had to be taken care of. So she was accompanied to the registration office by another woman, who introduced herself as the partner of the person who had brought her here, in order to register. She did not learn anything about a tax liability. She also had to sign a lease for the apartment. She was told that this was customary and that she should act as joint tenant of the flat. Accordingly, electricity and service charges for the flat were also in her name. As Ms. I did not speak German, she could not understand the contracts she signed. However, she understood that her name was written as debtor in all the reminders that arrived over time. She was put under pressure to work off these debts or she would face a prison sentence for fraud. Ms. I worked at all hours and always had to be available for potential clients. She was allowed to leave the house for short periods to go shopping, but had to sign in and out or was accompanied by other women. She had to give two thirds of the money she earned to the owner of the apartment, she was allowed to keep one third, but had to pay her living expenses and the costs for photos, advertising and working materials. At the beginning she was able to send about 200–300 € home every month via Western Union, but she saved on food and her personal needs. When she was ill she still had to pay the money for the rent, when she could not afford this she was beaten up by the apartment owner and his partner. She was told she had to earn enough money to pay her expenses. She had to receive more and more clients and provide more and more services, which she refused to do for herself. Sometimes she thought about running away and working somewhere completely on her own account. However, she was too afraid to get help and didn't know where to turn or how to explain her situation without knowing the language. The most important reason to stay in her situation, however, was the fact that although she had to give away most of the money she earned, she could send money home and these amounts were sometimes higher than her previous income.

After some time, Mrs. I asked to be allowed to travel to Bulgaria for a short time. Her father's illness and his health had deteriorated. She travelled back for two weeks with the same people she had come here with. At home she tried to make her family understand that she had to work in undignified conditions, but they demanded that she go back and continue to send money because of the family's economic deficit, otherwise her children could no longer be cared for. Ms. I travelled back to Germany and continued to work as before. She was picked up in the course of a police raid at her place of work and taken for questioning by the police. Here she did not give any information for fear of losing her only source of income, but was referred to the BBMeZ counselling centre. In a conversation with the counsellor and an interpreter, she gained so much confidence that she reported on her situation, but she decided not to take any further steps but to return to her old living conditions. The reasons she gave were the economic situation of her family and the fear of not being able to provide for her children, but also the pressure of her family's expectations, the lack of support to leave prostitution and fear of reprisals by the perpetrators. Later, there was further contact with the counselling centre. On one occasion a trip home was organised for her, on another Ms. I was about to press charges against her pimps because they had beaten her so severely that she had lost two teeth. She was temporarily taken to a safe place by the counselling centre, but she did not stay there, but went back to her old living conditions after a few days. Although she often felt unfairly treated by the perpetrators, she does not see herself as a victim of human trafficking and would like to continue trying to earn her money as independently as possible in the prostitution milieu. She is prepared to pay a high price in terms of self-determination in order to support her family. The limits in her activity, with which she no longer agrees, depend on the earnings and the share she is allowed to keep for herself and on how much she can and must endure for this.

Case Study 4: Ms A., Born 1983, Nationality Ukraine

Ms. A. has been living in Germany for about two years. Her home town is an industrial city in the Ukraine, where she worked as a labourer in a textile factory. She has completed an apprenticeship, but would have preferred to study after finishing school. Her unemployed parents asked her to contribute her share to the family income, so the desire to study was put on hold. Through her brother, she received an offer from one of his acquaintances to go to Germany as a service worker. Although she knew that she could not enter the country legally to take up employment, she was told that it was possible to apply for it locally. She decided to take the risk. She was told of good earning opportunities and contacts at casting agencies for modelling jobs she might be eligible for. A long journey ensued with three other women and several changes of vehicles through different countries. They were taken by boat at night across the Oder River to Germany, where they were met by several men. They were taken together to a house, assigned a dormitory, and over the next few days they were to meet their new employers. After a week she was instructed to get into a waiting car, they promised to take her to her workplace in a small restaurant. Once there, she was told that from now on she would have to work as a prostitute. She was beaten and raped. This was repeated in the following days until she agreed to become a prostitute. She worked in a night club as a waitress and prostitute for 1.5 years. She had no possibility to contact her home. She could not get help because of the lack of language skills and the great fear of the perpetrators. She hardly got any money for her work, it was charged by the perpetrators with the costs for her travel, accommodation, food and clothing. They had promised to pay her a large sum when she returned home, but until then it was safer to save the money for her. During a police raid, she was found in a prostitution establishment and taken away by the police for being there illegally. She was questioned about her situation and decided to testify, which she did in the following weeks. At present she is waiting as a witness for the forthcoming court proceedings. She is supervised by the counselling centre when needed, but only wants to have limited contact with counsellors. She has little trust in other people and wants to sort out her affairs herself as best she can. Until the start of the trial she will be the mother of a child, she is expecting a child from a German man and as a result will get permanent residence, which is independent of her witness status.

As the preceding case studies make clear, the origin and life situations of women in prostitution are very heterogeneous. There is no prototype of the typical (forced) prostitute. Each case has its own biographical history and is linked to the person who experienced it and to their personal life situation as well as their options for action and decisions. The women's perceptions and assessments of what happened to them are correspondingly different. The perceptions of the social workers who come into contact with women in prostitution in a counselling setting are just as different, and the intervention options they can offer are correspondingly different. There is no one-size-fits-all solution.

If one follows the Criminal Code and the aforementioned criminal offence of trafficking in human beings, there are certainly also clear factors by which forced prostitution and thus situations characterised by involuntariness can be recognised or legally classified and objectified. These in turn depend on the extent to which they can be substantiated and subsequently assessed by law enforcement agencies. This perspective, namely that of law enforcement, initially sounds quite clear, insofar as provability is given and the person concerned identifies with the legal classification. At least there are legal parameters that help with a classification. In practical social work with prostitutes, on the other hand, it is not always primarily a matter of assessing factors relevant to criminal law. "The most important goal is to support and advise them in their life situations so that they can lead a self-determined and responsible life" (Albert & Wege, 2011). It is the task of specialist counselling centres to assess their life situation and to clarify with them whether there is a desire for change and thus a need for action, and what framework for action is available. The legal dimension and the assessment of the situation from this point of view are not always the primary focus of those being counselled, but potential trafficked persons must always have the opportunity to learn about their rights during counselling. It is up to them to decide whether they want to make use of these rights.

5 Counselling Dilemmas in Social Work with Women in Prostitution and with Possible Trafficked Persons

Counsellors are repeatedly confronted with situations that take place in the grey area between voluntariness and coercion in different and sometimes rapidly changing manifestations. A clear classification and a clear positioning of the counsellor are not always possible and also not always desired by the client. Sometimes counselling situations arise in which the counsellor suspects, sometimes more or sometimes less clearly, that the client being counselled is affected by a personal predicament. Sometimes diffuse indications give rise to clues that suggest that self-determination is restricted. Sometimes the client herself has given indications of typical predicaments, which a counsellor then assigns on the basis of her wealth of experience. In order to be able to make this professionally sound assessment, it is necessary to have professional knowledge about living situations in prostitution, but also about milieu structures[9] and typical procedures of pimps, human traffickers and other participants, which counsellors acquire as professional tools in the course of their work. Furthermore, it is necessary to get a personal impression of the person to be counselled and to make an assessment of how they see and evaluate their situation. Sometimes it becomes clear that for outsiders the person's assessment patterns do not match what he or she reports in the counselling session. In order to do this, it is necessary to record as objectively as possible how exactly the living conditions present themselves. The following key questions can serve as a guide:

1. Does the client describe her prostitution activity as voluntary? If so, does she actually have other options?
2. What is its own framework for action?
3. Can she decide for herself when and where she works at any time?
4. Does it determine itself at what prices it offers which services and which services it does not offer?
5. Can she move freely? Does she have to justify herself to anyone if she wants to leave her prostitution site?
6. How much money does she pay for her place of work? Is this payment in proportion to what she earns and can realistically earn?
7. Does she have to pay for anything other than rent? If so, is the payment proportionate to the respective benefits?
8. Where does she live and sleep?
9. Does she have social contacts? Can these be described as trusting?
10. Does she herself feel comfortable or uncomfortable with her job?
11. How does she find her customers and how do her customers find her? Is she mediated by third parties?
12. Was she brought into prostitution by deception, false promises, or exploitation of a helplessness?

[9] Places, people, offers, etc.

13. In her situation, can she say at any time that she wants to get out and implement this without hindrance from others?
14. Is she being pressured physically or psychologically by others?

Not all of the questions listed above as examples that can be evaluated as possible indications of an activity that is not self-determined can always be answered unambiguously with a yes or no. There are also no mathematical or other systematic calculation schemes for this. Just as little are there mathematical or other systematic calculation schemes for this, within which it is clearly defined how many positive or negative answers clearly lead to an evaluation as a coercive or non-coercive situation. The consequence of this is that the assessment of a correspondingly unclear or changing life situation by a professional social worker cannot always be clearly assigned by the latter either.

At the same time, however, there is also the situation that a classification by the counsellor is very clear, but here the counsellor's professional assessment turns out to be different from how the counsellee sees him/herself. If there are many indications of a clear predicament with which the person concerned cannot identify, a discrepancy may arise for the counsellor between what is desired on the part of the counsellee and what appears appropriate from a professional point of view. This is not a new phenomenon in social work. However, in a field of work such as prostitution, in which there are so many and rapidly changing manifestations, which are accompanied by equally different personal assessments of the practitioners, this is problematic for counsellors when there are obviously inhumane working conditions. Counselling centres see themselves as a partisan point of contact for those affected and represent the clear maxim of not acting against the will and against the consent of those to be counselled.

Sometimes, however, counselling situations arise in which, from a professional point of view, it can be clearly classified that a person affected is being massively restricted and violated in her fundamental rights. It is possible that she cannot identify with this assessment for various reasons and rejects intervention by the counsellor, who, however, sees an urgent need for action. In this case, can action be taken against the client's will? Or must the decisions of the person concerned be respected and professional and presumably also personal concerns be set aside? The way out of this dilemma is quite simple if it becomes clear that there is a danger to others or to oneself, or that third persons are affected by the consequences of a decision against intervention. This can be, for example, a risk to the well-being of children, if children are present and the actions of a counselled person lead to the fact that their well-being is endangered and it becomes necessary to bring about an intervention by youth welfare offices.[10] It is also possible that a counsellee shows clear signs of suicidal tendencies.[11] In both cases, there is an immediate need for action against the will of the person being counselled. This can be clearly justified legally and morally by the aforementioned legal requirements. However, it is limited to the area of action that brings about a change in the acute danger and must not be understood in such a way that possible trafficking backgrounds are passed on to state agencies against the will of the person concerned. Intervention may only take place within the framework of the acute situation and is limited to remedying this dangerous situation. However, situations are not always so clear-cut, nor are the resulting consequences of action so clearly regulated. As a rule, it is at the professional discretion of counsellors to grasp and classify a situation. In the experience of the BBMeZ counselling centre, for example, it is not uncommon for clients to suddenly break off contact when counselling prostitutes. Often those affected who see a need for change only find their way and access to a specialist counselling centre when they have reached a point in their suffering where they can no longer find solutions for themselves and need outside help. In the case of trafficked persons who are or could become witnesses in criminal proceedings, for example, there is a risk of external influence and danger from circles of perpetrators, regardless of whether a person has already decided in favour of or against cooperating with the criminal investigation authorities. It is not always possible for a counsellor to clearly assess whether the break-off of contact has occurred voluntarily because the client has decided to do so or whether she may have been externally influenced by threats and/or the use of violence and is therefore no longer appearing in the counselling setting. At this point, the counselling person must make a decision as to whether to involve law enforcement authorities and thus disclose the life situation of the counselled person without her being able to actively participate in this decision and, if necessary, also decide against it. On the other hand, the counsellor runs the risk that a danger may arise of which he or she was aware as a counsellor and which could possibly pose a risk to life and limb for the person concerned. Here it must be weighed up very carefully and always on a case-by-case basis how to proceed, which concrete indicators are present and which measures are taken or not taken. In BBMeZ's advisory practice, there have been interactions in one direction as well as in the other. The respective decision to proceed in a certain direction was made depending on the previous history and its assessment in the case of women who had already had contact with the police. If a

[10] Cf. SGB VIII § 8a.

[11] The basis for the intervention is a law on assistance and protective measures in the case of mental illness; for Bremen, the PsychKG as of 13.12.2011 applies.

threat appeared to be concrete, the police were called in. In the case of women for whom no contact with the police was desired and the danger could not be concretely named, non-governmental agencies were informed in some[12] cases that might come into contact with them, for example the counselling centre for prostitutes or the public health department. Both services also work independently and anonymously. In the context of counselling, only once has a clear and concrete endangerment situation arisen in the case of persons who did not wish to have contact with the police, so that the counselling centre was compelled to take action against the wishes of the counsellee and involve the police. In this case it turned out that there had already been a Europe-wide search report on this case and that there was indeed a massive danger for the person concerned. Situations regularly arose that were unclear and where, on the basis of the previous history on the part of the counsellors, strong suspicions arose that there was outside influence and that the person concerned, under massive pressure on the part of the perpetrators, had supposedly decided on their own to return to a situation of coercion. In this case, it is usually not possible to speak of a self-determined decision, as factors from the victim's environment, such as the use of violence, threats or deception, have led her to decide to return and she sees no alternative course of action for herself. How can counsellors deal with such situations and handle them professionally?

6 Professional Handling and Positioning of Counsellors in Social Work with (Forced) Prostitutes

In order to be able to discuss the aforementioned question, it is first necessary to deal with the tasks of counselling centres and their mission statement in order to clarify the framework of actions within the counselling work. The KOK, as a nationwide coordinating group against human trafficking, has presented the self-image of the aid organisations united in the umbrella organisation, which are made up of specialist counselling centres for trafficked persons as well as for prostitutes and victims of violence in the migration process, as follows on its homepage:

- Specialist counselling centres are non-governmental counselling services for trafficked persons.
- Consultations are free of charge and anonymous

[12] Of course, in compliance with data protection and personal rights.

- mother-tongue counselling by staff with appropriate language skills or interpreters is possible
- the consultation is subject to professional secrecy, all information is treated confidentially
- nothing is done against the will of the advised (KOK, n.d.)

These preconditions have already been named in the previous section. However, it seems sensible to summarise these prerequisites under which counselling work for women takes place once again. The points mentioned must be reflected in the guiding principles for social work in the present topic complex and also as indispensable tasks for practical work. Only then can professional quality work be guaranteed. They form the basis of a trustworthy contact point for those affected, in which they can find orientation for their life situation and possible wishes for change. They are the theoretical superstructure and the basis of action for counsellors and thus their basis of legitimacy. It is never the task of an independent specialist counselling centre to act as an extended arm of the judiciary and its executive organs in order to prosecute criminal offences.[13] At first glance, it seems quite easy to consistently represent this position both within the counselling work and externally, since the framework and the terms of reference are clearly specified. In practice, however, new situations arise time and again in which the concept of voluntariness is interpreted and lived out in different ways, and counsellors are confronted with complex problem situations that they cannot always solve together with the persons concerned. For this purpose, it makes sense to look at two selected essential factors that can bring women into prostitution and here into difficult life situations, and to take a look at a possible way of dealing with them for counsellors.

Dealing with Cultural Foreignness and Other Gender Constructions

Again and again counsellors come into contact with people whose society of origin contains completely different evaluation systems and lifestyles. In particular, the

---

[13] Objections from the law enforcement sector that the state has an obligation to prosecute official offences such as human trafficking refer in this case to the criminal investigation authorities themselves. The latter are obliged to take action when they become aware of a criminal offence in this area. Independent counselling centres are not state officials and are therefore in no way obliged or entitled to pass on knowledge of a criminal offence without the consent of the person concerned.

position of the sexes in relation to each other and the different possibilities for action and development lead to the fact that primarily women have a clearly limited self-determination in comparison to men and does not correspond with the concept of equality of our social construct. As an example from the counselling work of recent years, the Roma minority[14] in many mainly Central and Eastern European countries can be cited, in which women have often experienced a role model that is inferior to men in terms of its status. There is a largely patriarchal social structure (European Parliament, 2013, p. 5). Women do not always have their own decision-making powers, i.e. self-determination as to how they want to shape their lives. In many cases, women are born into a family system in which their determination to shape their own lives is in the hands of primarily male family heads. It is not they who decide on their way of life and activities, but other persons or their families. This manifests itself in their upbringing and becomes part of their identity. Often they are shaped by experiences of violence and social and family pressure and are not familiar with making and reflecting on their own decisions. If they are confronted in the counselling work with the prevailing role models and the imbalance of their own situation, some women do not find themselves in the role of a person affected by coercive relationships. Some do not understand why actions and experiences that happen to them and have been part of their lives for a long time are not allowed here and cannot or do not want to recognise or accept a victim role for themselves. At this point the balancing act of a counsellor arises, on the one hand to point out the socially legitimised context, to grasp the life and origin situation and the counsellee and on the other hand to understand and accept the person and position of their counterpart and to look for possible solutions together with them. Counsellors need knowledge of these role constructs in the societies of origin in order to be able to offer individual solutions and thus take cultural characteristics into account. Above all, however, they need the willingness to come into contact with those affected on an equal footing and not to classify them as victims in the sense of helpless and will-less beings. Pointing out that assessments are different here in Germany, that certain actions are punishable and that those affected have rights is necessary and, of course, the task of a specialist counselling centre. However, this must always be done objectively and never in a judgmental manner. Only in this way can an appropriate offer of help be implemented in a helpful way for the person being counselled and in a professional way for the

[14] At this point it should be expressly pointed out that it is not a question of discrediting an ethnic group across the board, but rather experiences from the counselling work are to be listed, which have repeatedly become apparent after a thorough examination of the life situations of the counsellors.

counsellor. It is not always easy, from the perspective of a counsellor, to maintain acceptance of and a sense of empathy towards lifestyles that are judged to be unjust from the point of view of the society that exists here and its concepts of norms and values that they have internalised as counsellors and as human beings. Here it must always be made clear that the acceptance of otherness to the disadvantage of the person concerned does not apply to the system itself, within which she has grown up and whose values and norms she has internalised, but solely to her as a person. It is therefore about a personal encounter from person to person in a professional counselling context and without devaluing the other person. If it is possible to maintain this openness and not only to internalize the negative evaluations, there is a basic attitude of acceptance towards the counselled persons. If one succeeds in approaching them without any patterns of evaluation or devaluation, understanding their biography and also their role in their society of origin, personal incomprehension can be avoided and a professional approach can be manifested as a basis for work.

Dealing with Economic Inequality

In addition to the cultural backgrounds of many trafficked persons already mentioned, the aspect of female poverty migration plays an important role in the context of human trafficking. Economic constraints such as a lack of social security through a social assistance system accessible to all, a lack of educational opportunities and thus severely limited job opportunities, access to a viable health insurance system or low incomes are repeatedly the focus of trafficked persons and are listed in counselling as a motivation for migration to Germany. In many cases, the background to their lives is characterised by great poverty, existential hardship and a marginal existence. This leads to the decision to go to a foreign country and earn a living for the family in the home country (ERRC, 2011). Quite a few women have one or more child(ren) or sick family members at a young age whom they cannot take care of at home in their home country. Their financial resources are very limited and they see the opportunity of more lucrative work abroad for themselves and their families as a chance to lead a better life and escape poverty. They are often expected to take responsibility for the family's financial situation, but they rarely have power of disposal over the resources they earn (Jones, 1998). In addition to the aspect of cultural role identity, according to the experience of the BBMeZ counselling centre, economic constraints play a considerable role in the decision to take up an occupation in prostitution. This often takes place under false working conditions and is often accompanied by the promise of high profit rates, which suggest to those affected that they can solve their precarious financial situation through this. Even if only part of the money earned remains with them, this part is not infrequently higher than the income at home in the home country, which was previously used to make a living. Under this premise of massive economic pressure, more and more women decide to work in prostitution, especially since they often see no alternative in other fields of work due to a lack of education and stigmatisation. Knowledge of these conditions is very important for counsellors, because at this point general constraints such as poverty are linked with personal constraints of a family pressure situation.[15] It is not always possible to make those affected independent of this pressure situation, because many factors do not start at the personal level of those affected, but are the result of a social system, which cannot be changed through counselling work.

So what are the possibilities for improving the life situation of the counselled? In addition to strengthening the self-confidence of those affected to perceive themselves as persons capable of acting and as a gender with equal rights, the improvement of educational opportunities is an important approach. For example, in the work of the BBMeZ counselling centre, qualification in the area of reading, writing and language skills has certainly led in various cases in the work with women from prostitution to them reorienting themselves professionally and looking for work activities in which the danger of exploitation is lower and which nevertheless secure them an income with which they can feed their family. Remaining in prostitution is also a possibility, but under conditions that are characterized by working conditions that allow them self-determination and thus give them the chance to work for themselves and not for the profit of others. However, within the milieu, these conditions are not given to all women working there, and there is always a not inconsiderable risk of being affected again by exploitative conditions. Here it is the indispensable task of specialised counselling centres not to judge negatively the basic activity in prostitution, but to react appreciatively to this decision, provided that it was actually made self-determined by the counsellee. At the same time, however, it is necessary to address the risks and to work out with the clients what options they have to protect themselves from becoming victims of human trafficking again. By showing clients how and where they can get help and support if their perspective on a supposedly self-determined life situation changes, they do justice to their own professionalism as counsellors and can develop the feeling from their success in the work that they have made a constructive contribution to shaping the lives of the counselled.

[15] Which can arise from poverty and patriarchal social structures, as well as other factors.

7 Professional Devotion and Professional Demarcation

As already mentioned, there are different ways and approaches to professionally manage social work with women in prostitution and especially with trafficked persons. A prerequisite for access to a corresponding activity is always an appreciative attitude towards prostitution activities. If there is no personal acceptance in this respect, work in social work with prostitutes should not be considered, because this basic attitude will not be able to be concealed from those being advised in the long term and will certainly lead to a lack of understanding, conflict liability and inappropriate advice and suggestions in the counselling setting.

A sound knowledge of structures, processes and conditions in the prostitution milieu as well as the knowledge of life backgrounds described in the previous paragraph are particularly necessary in order to make professional assessments and provide appropriate assistance. This and the personal attitude to prostitution mentioned above can result in a professional approach. If the clients feel valued and understood, this is also reflected in their reactions to the counsellors and shows them that they are doing their job well. Accordingly, not only is appreciation and respect for the counsellee necessary, it is also important for counsellors to report back to them the meaningfulness of their work. From this they can gain personal job satisfaction. It is true that there are also situations in which counsellors feel that their counselling does not lead to changes in the life situations of the counsellees or is not accepted by them. At this point it should always be made clear that they can only make offers. The decision about a change and also about how it is implemented can only be made by the clients themselves.

Despite all the commitment, it is necessary to maintain a personal distance from the often very touching fates behind the counsellees in order to preserve one's own mental health. From the point of view of a counselling centre, it can only be recommended to establish professional support structures such as inter- and supervision, various methods of systemic work for case discussion such as the implementation of a Reflecting Team (Andersen, 1990) or Reframing (Bandler & Grinder, 1985) and regular meeting and feedback structures between the employees. This serves to reflect on the work, to get clarity and orientation and to relieve and strengthen each other.

Boundary transgressions by counsellees, who sometimes have difficulties in adhering to the framework of action offered to them and, for example, in keeping to certain rules, occur again and again in everyday counselling. Here it is necessary for a counselling centre and the staff to set clear boundaries and to communicate these sufficiently clearly in order to explain the limits of their work. For this purpose, there should be a consensus within the counselling centre, at best within the framework of its own conceptual design, on points of view, attitudes and possibilities for action, in order to provide the staff with a uniform basis for their work, which they can jointly represent to the outside world. Furthermore, employees in social counselling work should be aware that their own sphere of influence is limited and that they themselves cannot change an entire system which they may feel to be unjust and not in conformity with their own values and also with the prevailing social structures here. The offers of help for clients start at the individual level and can, as a rule, only lead to changes there at first. It is helpful to be aware that every small change for the better is a step in the right direction for those affected. Furthermore, counselling centres, as party representatives of the interests of a particular group, have the opportunity to draw attention to existing grievances through public relations, lobbying and awareness-raising work and to demand changes from those responsible for action, such as political representatives or other decision-makers. They give a voice to those affected and their needs, sensitise the public and can thus detach themselves from the personal level by formulating systemically anchored deficits and calling for remedial action.

8 Summary

This article is intended to show, at least to some extent, how complex and multi-layered social work with trafficked persons is and what challenges employees are confronted with in the context of social counselling. Fluent transitions between self-determination and coercion repeatedly present employees with new challenges in developing constructive approaches to solutions with trafficked persons. At the same time, counsellors have to put aside their own personal evaluation patterns and ensure an objective approach for themselves and the counsellee. This is not always easy and dilemmas often arise not only for the counsellee but also for the counsellor. It is a great challenge to endure this situation and still remain constructive and solution-oriented within the framework of a professional working method and also to accept that there is not an optimal solution approach from the counsellor's point of view for every situation in the life of a person concerned. It can happen that the assessment of a counsellee turns out to be completely different from the assessment of a specialist counsellor. Nevertheless, it is imperative to always approach each new counselling case in an unbiased, appreciative and objective manner.

As an independent counselling centre, it is not always easy to adopt an independent position and to defend it consistently in the often tense field of socio-political action. There is quickly the danger of becoming a political issue and being used for certain political purposes and goals. An example of this is the discussion that flares up again and again as to whether the legalization of prostitution has led to an increase in human trafficking. In this context, interest groups like to use the increasing number of cases in specialised counselling centres to show that a ban on prostitution is supposedly sensible and that this is the only way to effectively combat human trafficking. Contexts are presented in a grossly simplified way and serve to underpin certain political endeavours. The interests of the trafficked persons and their needs are seldom in the foreground and it is easily forgotten that approaches to solutions in this regard must be designed in a much more complex and differentiated manner than the mere demand for a ban. In the public perception, there is often a reduction to populist aspects and mono-causally structured explanatory patterns, so that one of the tasks of specialised counselling centres for trafficked persons is to publicly represent background knowledge, experiences from the work and needs oriented towards the trafficked persons as a lobby institution.

To clearly represent the position of an independent institution in the sense of a partisan work for the counselled is a task that requires a high degree of openness, acceptance and respect for other lifestyles and social concepts not only on the professional and institutional level, but also for counsellors on the personal level.

In this context, the question of compulsion and voluntariness can at best be answered by the prevailing system of values and norms, which manifests itself in the legal framework. Whether those affected can actually find themselves in this system varies from person to person.

References

Albert, M., & Wege, J. (2011). Soziale Arbeit und Prostitution. Handlungsbedarf und Entwicklungsmöglichkeiten in einem tabuisierten Berufsfeld. *Soziale Arbeit, 1*, 8–18.

Andersen, T. (1990). *Das reflektierende Team. Dialoge und Dialoge über Dialoge*. Modernes Leben.

Bandler, R., & Grinder, J. (1985). *Reframing. Ein ökologischer Ansatz in der Psychotherapie (NLP)*. Junfermann.

Bundesweiter Koordinierungskreis gegen Frauenhandel und Gewalt an Frauen im Migrationsprozess e. V. [KOK]. (Hrsg.). (2008). Bericht der Bundesregierung zu den Auswirkungen des Prostitutionsgesetzes – Anmerkungen und Empfehlungen für den KOK und seine Mitgliedsorganisationen. Berlin.

Bundesweiter Koordinierungskreis gegen Frauenhandel und Gewalt an Frauen im Migrationsprozess e. V. [KOK]. (Hrsg.). (2012). Handbuch zur Aus- und Fortbildung und Qualitätssicherung für Fachberatungsstellen für Betroffene von Frauen/Menschenhandel. Berlin.

Bundesweiter Koordinierungskreis gegen Frauenhandel und Gewalt an Frauen im Migrationsprozess e. V. [KOK] (Hrsg.) (n.d.). Hilfsangebote der Fachberatungsstellen. http://www.kok-buero.de/hilfsangebote.html. Zugegriffen 3. Jan. 2014.

Europäisches Parlament. (2013). Förderung von Roma-Frauen innerhalb des EU-Rahmens für nationale Strategien zur Integration der Roma. http://www.europarl.europa.eu/RegData/etudes/etudes/join/2013/493019/IPOL-FEMM_ET%282013%29493019%28SUM01%29_DE.pdf. Zugegriffen 4. Jan. 2014.

European Roma Rights Center [ERRC]. (Hrsg.). (2011). *Trafficking in Romani communities – Breaking the silence*. Fo-Szer Bt.

Hurrelmann, K. (1989). *Einführung in die Sozialisationstheorie*. Beltz.

Jones, A. (1998). Migration, ethnicity and conflict: Oxfam's experience of working with Romani communities in Tuzla, Bosnia-Hercogowina. In C. Sweetman (Hrsg.), *Gender and migration*. Oxfam.

Kalthegener, R. (2008). Rechtliche Grundlagen des Phänomens Frauenhandel zur sexuellen Ausbeutung. In KOK. (Hrsg.), *Frauenhandeln in Deutschland* (S. 35–48). Berlin.

Schmidt-Salomon, M. (1995). Können wir wollen, was wir wollen? Unzeitgemäßes zur Theorie der Willensfreiheit [Elektronische Version]. *Aufklärung und Kritik. Zeitschrift für humanistische Philosophie und freies Denken*, 2.

Schoppenhauer, A. (1977). Preisschrift über die Freiheit des Willens. In A. Schoppenhauer (Hrsg.), *Werke in zehn Bänden* (Band VI, S. 48). Diogenes.

Spittler, G. (1967). *Norm und Sanktion*. Walter.

Katharina Kähler works at the counselling centre for trafficked persons and forced prostitution (BBMeZ) of the Inner Mission in Bremen. The BBMeZ helps women who have become victims of human trafficking and/or forced prostitution.

Coaching, Supervision and Process Support of Counselling Centres for Sex Workers

Angela Siebold

1 Introduction

In this article, the central themes from supervision and coaching processes are presented and analysed in terms of their significance for the further professionalisation of counselling centres for sex workers. These include topics that arise from the encounter with the sex workers' field of work as well as topics that arise in the different spaces of interaction and communication. In addition, topics are relevant that are important in the discussion of individual experiences and reflections on the professional role of the social worker/counsellor in a professional field of work that has not yet been adequately clarified by legislation and tends to be given a negative image by society. Under the concept of process accompaniment, a concept of comprehensive accompaniment in the personnel and organisational development processes of the specialist counselling centres, integrated in the social context, is presented, which at the same time also enables outlooks into future developments and task fields of all institutions, organisations and model projects active in this professional field.

A. Siebold (✉)
Siebold, Germany
e-mail: siebold@siebold-coaching.de

M. Albert, J. Wege (eds.), *Social Work and Prostitution*,
https://doi.org/10.1007/978-3-658-37761-8_12

2 Understanding of Coaching, Supervision and Process Support

The following article deals with the importance of coaching, supervision and process support in counselling centres for sex workers. It refers to counselling centres which, as members of the Federal Association of Specialist Counselling Centres for Sex Workers (Bufas e. V.), advocate for the rights of sex workers, demand their participation in all decisions affecting them and also claim permanent funding for a counselling infrastructure for this target group. Many of the professional counselling centres united in Bufas e. V. were formed in the 1980s or 90s as part of the new social movements and see themselves as a "whore movement". Their roots can be found in the women's movement and in self-help and self-organisation (cf. among others the concepts of Madonna e. V. in Bochum, Hydra e. V. in Berlin, Kassandra e. V. in Nuremberg). Thus, from the very beginning, the construct of empowerment (Herriger, 2010) was manifested in all concepts and forms of organisation. Sex workers, former sex workers and social workers had joined together in associations with the essential goals of improving working conditions in sex work, their professional and social equality and improving the living situation of sex workers. From the beginning, counselling according to the principle of the women's movement "sex workers advise sex workers" was integrated into the self-help movement and developed in the 1990s into a differentiated concept of psychosocial counselling in interdisciplinary specialist counselling centres in connection with debt counselling, health prevention and qualification offers. In the meantime, counselling centres have also developed under church sponsorship or within the framework of public administration, e.g. in the health offices. In some cases, very intensive cooperation has been established between these different types of organisations. For example, employees of the public health department are ideally or structurally (e.g. when they take over a board function) connected with counselling centres as members of the association, or there is joint work in outreach health prevention.

The concepts are often based on Thiersch's lifeworld orientation approach (Thiersch, 1992; Grunwald & Thiersch, 2008) with outreach counselling at the various locations of sex work and with counselling, information and qualification services oriented towards the lifeworld of the women in a counselling centre. "Helping women to help themselves" is usually the overriding postulate here. This usually refers to the concept of empowerment, even if this is not always so explicitly clear to all those involved, which also indicates the first potential for conflict, which is one of the central topics of supervision.

In the following, the most important topics from supervision are presented, here understood as team supervision or team development (Schreyögg, 2010, 2003), in order to then present the area of coaching for managers and finally to show how new fields of action can be developed in organisational development processes as well as solutions for current and future challenges in the counselling centres.

On the question of the distinction between supervision, coaching and process support, the professional association of supervisors, the German Society for Supervision (DGSv), formulates in a statement that the distinction is not professionally decided and that the attempt to distinguish between the terms would mean answering a "question that is undecidable in principle" (von Foerster, 2003) (DGSv, 2011). Both consulting formats are aimed at organisations and their members and accompany them in coping with change and development processes in the modern world of work (DGSv, 2011; Buer, 2012). In both counselling formats, both the content-related topics and the process orientation as well as the multidimensional view of persons, systems and contexts are interconnected. Nevertheless, this paper is oriented towards the differentiation of the two forms of counselling according to Schreyögg's (2003) proposal, as it is helpful as a construct to represent the complex events of accompanying developmental processes in counselling centres for sex workers. For example, Schreyögg sees supervision as working on factual issues and the interactions between team members as accompanying "change from below" and as "personal development". Coaching, on the other hand, is the counselling of managers, as "change from above" and as "personal development", focusing on issues of management and leadership (op. cit.). In process coaching, both processes are then linked in a comprehensive learning and development process, which also includes intentional learning processes and competence development processes, such as training on intercultural action competence, leadership training or communication training, or formats such as workshops on the development of a common vision or a quality policy, etc. Process orientation means that all these settings and methods are mutually dependent and complementary in a development process that enables the organisation and the people working in it to face the constant changes and challenges in social work.

3 Central Topics in Supervision

One of the most important concerns and often the reason for the request for supervision is the diversity and its effects on the cooperation in the team. Conflict potential lies in the interdisciplinary cooperation of (former) sex worker, social worker, psychologist, cultural mediator or multiplier, in the different integration into the counselling centre and the respective roles of the employees in the organisation (full-time, voluntary or part-time, permanently employed, freelance or a precarious employment relationship). Supervision is about understanding the complexity and the manifold, interdependent relationships, about central values, about dealing with hierarchy and power, about competition and about the shaping of professional relationships. It is necessary to deal with the topic of sexuality and to develop an individual and a common attitude towards sex work. Here, the tense term pair "solid woman – sex worker" is often mentioned, not infrequently with the insinuating assumption that sex workers are per se on an equal footing with the women to be counselled. Non-sex workers feel pressured to first prove their accepting stance. Further competition arises from different disciplines, socialisation backgrounds, employment conditions, role profiles, etc.

In the supervised reflection on these topics, the team women develop an appreciative and respectful way of dealing with each other, describe their mutual expectations of each other, develop trust in each other and develop a common professional ethical attitude towards sex work and the milieu. This appreciative attitude, which clients claim as a matter of course, must first be developed for oneself and in the encounter with colleagues. Only on this common basis can it be defined what, for example, openness and respect towards the sex workers to be counselled means. The perception of self-deprecation, self-rejection or contradictions in the behaviour of clients can only be addressed when a trusting, appreciative atmosphere has developed in the team. In this process, the urgently needed tolerance of ambiguity is further developed, as an important prerequisite for being able to meet the client's own, sometimes incomprehensible decisions with openness and respect for her obstinacy. It is also a matter of developing common terminology and definitions, and of dealing with the underlying attributions and horizons of expectation, such as sex worker or prostitute, client or customer, client or john, etc. The discussion of experiences of discrimination is an important prerequisite for this. Dealing with experiences of discrimination, labelling, stereotypes and prejudices can also raise awareness of how to deal with racist attitudes and statements and promote the development of intercultural competence. In this context, there is also the development of a common understanding of partiality (Herriger, 2010, p. 38), a central concept from the history of the emergence of the counselling centres from the whore movement, which also has a high significance for the concept of empowerment. At the same time, this concept requires an examination of the different interpretations and formations of the professional helper role. What is the effect of the asymmetry of the relationship between counsellor and client, how transparently is power dealt with, what does self-empowerment mean in the sense of the empowerment concept, how do we understand our self-imposed mandate, how does this relate to the performance of publicly funded tasks such as prevention work on street prostitution, in cooperation with other institutions, especially when it possibly contradicts clandestine goals of these? (cf. Vorheyer & Nagel, 2011). Especially in cooperation contexts with female employees from public institutions, considerable entanglements can arise here, as Vorheyer and Nagel (2011) impressively describe in their essay "Dealing habitually with the paradoxes of professional action: social work in the prostitution scene". The conflict presented here between the concern for empowerment, the acceptance of the client's obstinacy and the (clandestine) mandates such as control and integration into the public help system, between committed partiality and service orientation (Herriger, 2010, p. 66 ff.) is often resolved in an exclusive "practice focus". "Doing and making determines the goodness of the work, not how things are defined, but how they are shaped in the here and now of the concrete situation" (Vorheyer & Nagel, 2011, p. 20). There (...) "seems to be a desire not to reflect to the last consequence the antinomies between sovereign control behaviour of the administration and professional, advocatory social work" (op. cit., p. 25).

Herriger describes further potential for entanglement that arises from the provision of public funding and an implied, partly open, partly secret expectation. The staff members have to place individual problem presentations of the clients "(...) in institutionally and professionally recognized problem drawers" (Herriger, 2010, p. 66), "(...) for which institutionally pre-programmed intervention procedures are available" (loc. Cit.). This "preparation" involves the danger that the operation of deficit-oriented systems of assistance can also be associated with attributions that run counter to an empowerment approach. Likewise, the system logic of public administration and public funding or subsidy providers, which are to be outlined here with the keywords service orientation, formulation of help needs and target agreements, documentation obligation, has an effect both on the cooperation with colleagues from these systems as well as on one's own work. These dependencies, entanglements and contradictions must be revealed in supervision in order to clarify their effects on the perception of the professional role and to develop a professional self-image in recognition of them.

The same applies to the "interpretation foils" (Herriger, 2010, p. 67) that result from the practice orientation. These interpretation foils from the "everyday theoretical professional knowledge" (loc. cit.) must be revealed in supervision in order to uncover the view of existing resources in the client system and to relate them to an empowerment approach that is to be understood holistically. This means, in addition to the individual case processing, also discovering the resources and self-control and self-organisation possibilities both in the environment of the client system and at the level of the peer group, the milieu, the neighbourhood, the neighbourhood and the counselling centre (Herriger, 2010).

In this context, the questions of demarcation and professional distance also arise. Encounters and feelings from the client system such as powerlessness, resignation, heteronomy, disgust, self-deprecation, helplessness, etc. are reflected at the level of self-perception in the professional role and in communication and cooperation in the team and are cognitively and emotionally processed in supervision. The same applies to transference and counter-transference phenomena from the counselling situations. Together we can work on the further development of professional competences, stabilise a self-image in the perception of professional roles and learn to develop new possibilities of empowerment strategies in cooperation and from each other. In this way, the role of the helper can be shaped more intensively as an enabler, as a resource discoverer and as an optimistic companion. This further development of the professional role from helping expert to companion and facilitator with expertise requires courage and a deep trust in the client systems, in oneself and in the colleagues in the team, which has to be re-established again and again. The setting of supervision is the appropriate field of discovery and learning for this development. However, this presupposes that the supervisor herself is shaped by this inner trusting and optimistic attitude and, in particular, has reflected on her own ethical stance on sex work and can approach the subject with credible appreciation. With this understanding of the professional role, the topic of dealing with stress and strain can also be dealt with. It is acknowledged that sex work is a profession with very special psychological and physical stresses and high risks (Brückner & Oppenheimer, 2009; Schrader, 2012). Both the encounter with these problematic situations, with health hazards, violence, drug use, exploitative relationships and oppression is stressful for the female employees as well as the constant feeling of "not having done enough" in the face of great distress or not being able to record any successes because the counselled woman earns or has to earn her money elsewhere in the meantime (Winter, 2009, p. 223). Likewise, the constantly changing situations in prostitution (new women all the time, strong increase in migrants, especially from Eastern European countries, changes in the milieu, etc.) (Winter, 2009; Howe, 2012) can have a burdening effect, as can the overall conditions in prostitution, which are still very unsettling despite the Prostitution Act, which can have a detrimental effect on the work of the counselling centres. In addition, there is the confrontation with a diffuse social image of prostitution and the increasing public discussion, also in Germany, in which the topics of human rights violations through human trafficking, poverty migration movements in Europe and prostitution are linked together in a partly inadmissible way by appealing to a moral view of a part of the feminist movement.

> The scale in the field of prostitution ranges from high respect to complete contempt. The topic obviously triggers a deep-seated unease and is characterised by a stigmatising "half-knowledge". This circumstance must also be taken into account in practical social work. (Howe, 2012, p. 37)

This affects each individual counsellor in the exercise of her professional activity in contact with other institutions, in the public relations work of the counselling centres and in the lobbying work of the associations, as well as each counsellor in her immediate personal environment. With the label of the paternalistic role of the social worker who helps the victims of violence and human trafficking, she receives recognition and sometimes "admiration" ("I couldn't do that"). With the approach of "helping people to help themselves" she can also still count on the recognition of professionals from other areas of social work. With an emancipatory understanding of the self-help approach in the sense of the empowerment concept, with the advocacy of the rights of women in prostitution, the recognition of the professional activity of sex work and the corresponding improvement of the working conditions in prostitution, she is more likely to experience rejection in a social environment that is very much shaped by a debate in the media about prostitution in inevitable connection with human trafficking, crime and the degradation of women.

The danger of exclusion, self-exclusion and (professional) isolation that exists here due to one's own involvement as a sex worker in the counselling role, due to close connections in the milieu and its own structures and cultures, and due to the lack of or insufficient appreciation of the field of work "social work in prostitution" can be counteracted in supervision by making these phenomena visible and consciously noticing mirroring effects, so that the development of relieving counter-strategies is made possible as a result.

A further risk for the psychological stress of the female employees in the counselling centres lies in the financial and material framework conditions and their effects on the working conditions in the counselling centres. These include first of all the inadequate basic funding, which has to be supplemented by constant project funding and the solicitation of donations, and the associated job insecurity. The frequently inadequate premises for the multitude of tasks to be accomplished and thus the lack of space and forced flexibility also have a detrimental effect on the employees. In addition, in many cases the counsellor has to take on tasks that do not necessarily belong to her original performance profile, which can also have a psychological impact (Semmer et al., 2013).[1] This includes tasks that clients are expected to do even though they could actually do them themselves, as well as the completion of some bureaucratic files, documentation requirements and applications that are perceived as unnecessary. In the characteristics of "flexibility, uncertainty, project work, lack of time and increasingly fragile social relationships" (Voß et al., 2013, p. 67), occupational psychologists see possible risk factors for psychosocial stress, which in the worst case can lead to a so-called burnout syndrome (loc. cit., p. 65). According to the above-mentioned systematization by Schreyögg (2003), improving the framework conditions of social work in prostitution belongs to the area of responsibility of management and will therefore be dealt with in the following under the title of coaching.

4 Central Topics from the Coaching for Managers in the Counselling Centres

First of all, the question must be answered as to who actually belongs to the management in the counselling centres and how this is then structured, profiled and integrated into possible further institutional structures. This is where there is potential for conflict between voluntary board members and full-time social workers/psychologists in the perception of the management role (Schmidbauer, 2009). If there are (former) sex workers on the board (whose equal participation is included in some statutes of counselling centres in the tradition of the whore movement), the complexity of the issues and role perceptions can lead to unfavourable constellations and entanglements in which conflicts are personalised at the management level. Not infrequently, mutual mobbing processes even take place here, the negative and sometimes very destructive effects of which have to be worked through in team supervision and coaching processes for the management in elaborate conflict consultations.[2]

In addition to the central topics from supervision described above, the topics of power, perception of the employer role, cooperation between full-time and honorary positions, delegation of responsibility, and empowerment in the sense of volunteer lobbying and political strategies must therefore be dealt with at the management level. Once the common values have been worked out and mutual

[1] In the article, "Psychological Stress from Illegitimate Tasks," Semmer et al. (2013) describe how illegitimate tasks, which the employee perceives as unnecessary or unreasonable, increase stress because they threaten self-esteem.

[2] An example of these complex relationships is described by Multhaupt-Meckel (1997).

expectations clarified and defined in task profiles, descriptions of work processes and delegation procedures, the management is concerned with how the empowerment approach can also be reflected in the structures of the organisation and implemented with a participative management style. In this context, appreciative communication, the design of an appropriate organisational culture and an appreciative ambience are of central importance. The stresses of the work itself (sex work and counselling and support in the milieu) and the associated risk factors for the psychosocial health of sex workers and counsellors must be countered by stable, binding and reliable structures that enable trusting relationships at the collegial level and at the counsellor-client level. The chaos from the client systems must be contrasted with an order in the organisational system that makes it possible to recognise and structure the complex problem contexts. This includes a clear distribution of tasks, transparent decision-making structures, a structured information system, sensible time management that provides sufficient time for reflection and collegial consultation, as well as binding communication rules that ensure appreciative communication and prevent the danger that the rough communication style of the different milieus is reflected in the communication and cooperation structures and in the communication behaviour in the counselling centre.

In addition to this field of action of shaping the internal organisational framework conditions for the successful implementation of social work in prostitution, coaching also deals with other management tasks such as future development, financial security and the material equipment of the counselling centres, as well as personnel development and personnel management. In addition, the shaping of cooperative relationships and the performance of representative tasks in committees, such as the frequently established "round tables on prostitution", political lobbying and participation in professional associations are of great importance, as are marketing and public relations. In addition to efforts to permanently improve the financial and material conditions for the counselling centres, securing the future also includes shaping demographic change. The strong internationalisation of the sex industry (more than 50% of women in prostitution are migrants)[3] requires a comprehensive management of this diversity and differences in the different working areas of the counselling centres. As a rule, addressing the target groups can only be achieved through cooperation with native-speaking colleagues or so-called cultural mediators. This requires the development and implementation of a

[3] There is no reliable statistical data on the exact number of prostitutes in Germany. According to the Federal Ministry for Family Affairs, Senior Citizens, Women and Youth, more than 50% of prostitutes are of foreign origin (BMFSFJ, 2014).

language mediation concept compatible with their own approach to empowerment, such as the "Swedish model"[4] (Dhawan et al., 1995), as well as the sensible integration of these employees into the organisation. At the same time, the services offered by the counselling centres must react flexibly to the changing demand and, for example, provide language and literacy opportunities or offer sexual education in the respective mother tongue as well as promote and further develop the intercultural competence of all female employees. Another issue of demographic change is the shortage of skilled workers in the social economy, both in quantitative and qualitative terms (cf. among others an empirical study by Aquinet Business Consulting, 2012). In order to be able to compete with other fields of social work, the topics of maintaining health and strengthening the resilience of employees, reconciling family and work, personal and professional development opportunities, as well as an appreciative organisational culture for retaining employees are gaining great importance. In some "old" counselling centres from the whore movement, the preparation of a suitable successor and the development of intergenerational cooperation are also on the agenda.

The coaching is concerned both with the content of these topics and with how the manager presents himself as a leader both internally and externally, how he asserts himself and contributes to the development of the image of the counselling centre, and which management tools and methods are suitable to support his individual perception of the leadership role.

5 Central Topics for the Monitoring of the Processes for the Development of Future Fields of Activity in the Counselling Centres

Process support is to be understood here as support in the linking of various processes in a holistic learning and development process of the organisation to cope with future challenges. In view of the marginalisation, discrimination and extensive tabooing of the professional activity of sex workers (Kavemann, 2009, p. 178) and the exclusion of the topic of prostitution in related fields of social work (Hniopek, 2012, p. 90 ff.), the target group-specific and lifeworld-oriented approach will continue to be necessary and useful. However, it should be linked more

[4]The "Swedish model" in this context means translating according to certain criteria developed in Sweden: literally 1:1, neutrality and adopting an appropriate professional, detached attitude to translation, i.e. friends, relatives or colleagues cannot take on the role of translator because of the risk of role confusion and colouring of the language mediation.

intensively than before both conceptually with a comprehensive understanding of the empowerment approach and expand the connectivity to other systems of social work and education systems. The bridges to other supporting systems will only be sustainable, however, if it is possible to counteract corresponding discriminatory attributions and exclusions, which in turn requires empowerment strategies in political, social and professional discourses of social work. This is where the circle closes, which is why the following will focus on the development of an overall concept of empowering approaches and show how the potential of counselling centres can be further developed here. The approach of empowerment on all levels described by Herriger (2010) serves to structure this.

1. **On the level of the person,** this concerns the empowerment of the sex worker to cope with her life situation in prostitution and in the transition or exit phase. The previous socio-pedagogical counselling approach of "helping people to help themselves" with a strong practical orientation and the associated potential for entrapment (see above) can be supplemented by the solution-focused counselling approach (de Shazer, 1996; Bamberger, 2010). This approach supports the strong practical orientation of the employees, but reduces the potential for entanglement by focusing on the resources, potentials and action alternatives and can help to avoid the preparation for problem drawers with this consistent solution orientation. At the same time, the counsellors' overall perspective and attitude can change with effects on their language behaviour, so that appreciative communication in the team can also develop further with positive effects on the organisational culture. For this purpose, all employees of the counselling centre must be trained accordingly and the counselling concept must be expanded in this sense.
2. **At the level of the peer group,** the results of the ANAKO project and the practical experience of other projects, such as Pro Frieda (Diakonisches Werk der Evangelischen Kirche von Westfalen, 2007),[5] can be used to expand further education in the group in connection with consolidating their self-esteem and strengthening the relationships among the women. This education should be

[5] With the ANAKO project – Analysis of the Competence Profiles of Prostitutes as a Prerequisite for Socio-Professional Integration into the Primary Labour Market – Beate Leopold, Maja Czajka and Angela Siebold were able to prove that the activity in prostitution has essential characteristics of a profession, which form a good basis for possible careers outside of prostitution and that there is a very high motivation among the sex workers to obtain professional qualifications. Methodologically, it was possible to prove that an attitude which does not regard the women as "objects of social work, but as consciously acting subjects" (ANAKO final report p. 4.) forms a fundamental prerequisite for this motivation.

modular and embedded in an environment conducive to learning, by which is meant both the design of the counselling centre as a place of informal learning and the counsellor herself as a model. The processes in the counselling centre and in the training rooms can also be linked with many opportunities for informal learning that make self-efficacy tangible, evoke existing competencies and show mutual appreciation.

Building on the first positive experiences with mobile workshops at the place of work of sex workers by Stefanie Klee (Klee, 2008), new concepts of mobile learning offers can be developed, which can be integrated in this way into the everyday work in the brothel or on the street and at the same time strengthen the interaction of the women with each other. Here, the counselling centres have great potential through many years of experience in qualification, which can be expanded in the development of new conceptual approaches in model projects, but also in the consistent implementation of previous successful approaches.

A second important qualification approach is further training in the field of sex work itself in cooperation with institutions that are already established in this field, such as Pro Familia, health offices, self-help groups from the health sector or disability aid. Further training as a sex worker (or as a multiplier in health prevention) can not only further develop existing competencies and have an individual self-enhancing effect, but also at the same time contribute to destigmatisation and recognition of the professional activity with an enhancing effect on the group of sex workers as a whole.

Another possibility of self-empowerment lies in a systemic, solution-oriented supervision offer for sex workers, which at the same time also represents an important burnout prophylaxis. The experiences from the ANAKO project (Leopold et al., 2001) show that sex workers have a high motivation to learn for educational formats that are also carried out in other professions and that their professional activity is thus classified as equivalent. In this sense, the ZR concept, the Zurich Resource Model by Storch and Krause, which has shown great success in both career development and youth services, can also be applied for use in counselling and educational work for sex workers (Storch & Krause, 2011). The adaptation of successful approaches from social work with migrant women should also be critically examined and tailored to the needs of women in prostitution, such as literacy courses, or the concept of the project "With Migrants for Migrants – Intercultural Health in Germany ("Mimi")" from the Ethno-Medizinisches Zentrum e. V. in Hanover, which is designed to strengthen personal responsibility. Here, too, cooperation with migrant self-organisations, multicultural centres or adult education centres, etc., which

already have experience with this empowerment-based working approach, can be examined.
3. **At the level of the organisation,** all the topics can be found that were presented above under the concept of coaching for management and executives and are therefore only mentioned again for the sake of completeness.
4. **At the level of the neighbourhood and the social space,** Howe (2012) has shown with the research project "Neighbourhoods and Street Prostitution" that it is possible to develop solution scenarios for a peaceful neighbourhood, to improve communication and to increase mutual acceptance with activating, participatory and creative methods (participation of all people living and working in the social space, including representatives of regulatory authorities). From these positive experiences, concepts can be developed which the counselling centres implement in cooperation with other institutions located in the social space and link to their respective offers.
5. **At the level of society,** beyond the previous activities of the counselling centres (involvement in the round tables on prostitution, participation in committees and working groups, e.g. to improve the prostitution law, international cooperation, to name just a few of the numerous activities in the area of lobbying, press and public relations work), new topic-specific and/or strategic cooperation and connections can be entered into or deepened. For example, on the topic of sexuality, this could be a cooperation with Pro Familia, in order to bring the topic of dealing with sexuality and the value of sexuality more into the social focus, or with the trade union Ver.di, in order to push the topic of "labour rights for sex workers". It is equally necessary to set in motion an "alternative narrative, (...) a permanent counter-reading, to organize a positive occupation of the topic in order to break down the stigmas (...)" (Howe, 2013). Such "counter-readings" serve to reduce fear of contact, bring the topic of prostitution out of the dark field of criminalisation and taboo into a middle-class milieu and enable contacts with the economy and potential employers. These "counter-readings" form the framework for encouraging and empowering activities and counselling approaches in the counselling centres.

In order to be able to develop and implement all these approaches, in addition to the political and legal strategies described elsewhere in this book for improving the living and working situation of women in sex work, there is a need for sufficient financial, material and staffing resources for the counselling centres, as well as the establishment of counselling centres and round tables across as much of the country as possible to integrate and network all actors (police, public order office, building regulations office, trade inspectorate, tax office, counselling centres, equal opportunities offices, anti-discrimination offices, foreigners authority, employment agency, etc.). Even though there is certainly still a need for research in various places (e.g. in the target group-specific testing of the above-mentioned concepts of solution-focused counselling, CRM, mobile workshops, etc.), conclusions can first be drawn from the existing findings and corresponding measures that have been successfully tested in various model projects can be implemented and formulated as a quality standard for counselling centres in social work in prostitution. Process accompaniment thus takes place in the context of a complex social situation, as Schiersmann's (2009) "systemic context model of counselling" makes clear (cf. Fig. 1). The task of external process guidance is to make these complex contexts clear again and again and to raise the awareness of all actors involved.

In addition to these professional and content-related tasks, further challenges of external process support lie in the support of the development and implementation of an appropriate quality management system: "(...) a profession that wants to remain credible in the long term must provide information about what it does, why it does it and what benefits its actions are intended to generate" (Merchel, 2010, p. 32). The specialist advice centres provide this information in numerous reports

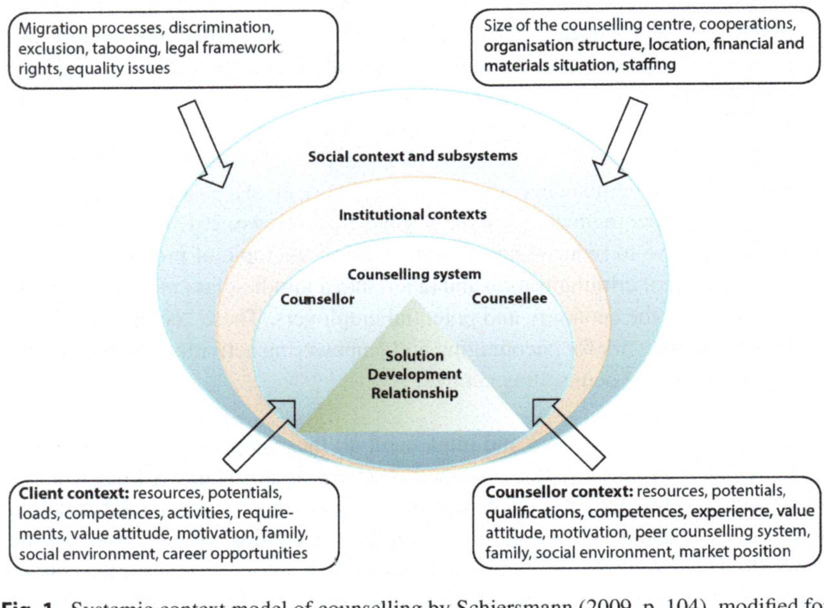

Fig. 1 Systemic context model of counselling by Schiersmann (2009, p. 104), modified for practice in specialist counselling centres for sex workers. (Own representation)

as evidence of their activities for public funding bodies or in documentation of funded model projects. As already mentioned elsewhere, however, there is a lack of systematic processing of all findings and experiences from projects and concepts and their theoretical foundation, which could then lead to the development of quality standards in the specialist counselling centres.

A quality management system based on the Total Quality Management philosophy, as it is now established in many areas of social work (Merchel, 2010) with a consistent customer or user orientation can address many of the issues raised in this article, make them accessible to systematic recording and further development, provide evidence of the effectiveness of measures and support services and thus lead to a permanent and comprehensive infrastructure of care, support, education and counselling opportunities.

References

Aquinet business consulting GmbH. (Hrsg.). (2012). *Fachkräftemangel in der Sozialwirtschaft. Eine empirische Untersuchung*. Hamburg.

Bamberger, G. (2010). *Lösungsorientierte Beratung* (4. Aufl.). Beltz.

Brückner, M., & Oppenheimer, C. (2009). Gewalt in der Prostitution – Untersuchung zu Sicherheit, Gesundheit und sozialen Hilfen. In B. Kavemann & H. Rabe (Hrsg.), *Das Prostitutionsgesetz. Aktuelle Forschungsergebnisse, Umsetzung und Weiterentwicklung* (S. 153–166). Barbara Budrich.

Buer, F. (2012). Die Supervision und ihre Nachbarformate. Was soll, was kann und was sollte das Besondere an der Supervision sein? In H. Pühl (Hrsg.), *Handbuch der Supervision 3* (2. Aufl.). Leutner.

Bundesministerium für Familie, Senioren, Frauen und Jugend (BMFSFJ) (2014). Prostitution. http://www.bmfsfj.de/BMFSFJ/gleichstellung,did=97952.html. Zugegriffen 7. Feb. 2015.

De Shazer, S. (1996). *Worte waren ursprünglich Zauber: lösungsorientierte Kurztherapie in Theorie und Praxis*. Modernes Lernen.

Deutsche Gesellschaft für Supervision e. V. (2011). Das Ende eines unerklärlichen Unterschieds. http://www.dgsv.de/2011/10/supervision-und-coachingganz-auf-einer-linie/. Zugegriffen 18. Juni 2013.

Dhawan, S., Entrena, E., Eriksson-Söder, U., & Landahl, M. J. (1995). Der Dolmetscher als Brücke zwischen Kulturen und Sprachen. In K. Peltzer, et al. (Hrsg.), *Gewalt und Trauma – Psychopathologie und Behandlung im Kontext von Flüchtlingen und Opfern organisierter Gewalt* (S. 178–192). IKO.

Diakonisches Werk der Evangelischen Kirche von Westfalen. (2007). *profrida – zeig deine Stärken – Auswertung des Modellprojekts*. Kurzfassung.

Grunwald, K., & Thiersch, H. (Hrsg.). (2008). *Praxis Lebensweltorientierter Sozialer Arbeit. Handlungszugänge und Methoden in unterschiedlichen Arbeitsfeldern* (2. Aufl.). Juventa.

Herriger, N. (2010). *Empowerment in der Sozialen Arbeit. Eine Einführung* (4. Aufl.). Kohlhammer.

Hniopek, A. (2012). Obdachlose Prostituierte – Klientel der Wohnungslosenhilfe? Überlappung, Überlagerung oder integraler Bestandteil? *Standpunkt: Sozial, 3,* Hamburger Forum für Soziale Arbeit und Gesundheit, S. 90–93.

Howe, C. (2012). Struktureller Wandel in der Prostitution. Zwischen Hurenbewegung und sozialer Arbeit. Standpunkt: Sozial, 3, Hamburger Forum für Soziale Arbeit und Gesundheit, S. 35–47.

Howe, C. (2013). Nachbarschaften und Straßenprostitution. Vortrag auf der Fachtagung 2012 des Bundesverbands der Fachberatungsstellen in der Sexarbeit (Bufas e. V.). http://madonna-ev.de/DOKUMENTE/Anhang%2015_Christiane%20Howe_Berlin.pdf. Zugegriffen 14. Juli 2014.

Kavemann, B. (2009). Auswirkungen des Prostitutionsgesetzes auf Ausstiegshilfen aus der Prostitution, In B. Kavemannn & H. Rabe (Hrsg.), *Das Prostitutionsgesetz. AktuelleForschungsergebnisse, Umsetzung und Weiterentwicklung* (S. 167–202). Budrich.

Klee, S. (2008). Modellprojekt: "Fortbildung für Sexarbeiterinnen in Clubs und Bordellen". http://www.highlights-berlin.de/DAH-Konzept.pdf. Zugegriffen 14. Juli 2014.

Leopold, B., Czajka, M., & Siebold, A. (2001). *Analysen der Kompetenzprofile von Prostituierten als Voraussetzung für die sozio-professionelle Integration in den ersten Arbeitsmarkt* (Anako). Wissenschaftlicher Abschlussbericht, Madonna Beratung und Hilfe für Prostituierte (Hrsg.). Bochum.

Merchel, J. (2010). *Qualitätsmanagement in der Sozialen Arbeit. Eine Einführung*. Juventa.

Multhaupt-Meckel, A. (1997). Supervision in einem Prostituiertenprojekt. Keine lustvolle Erfahrung. *Forum Supervision, 5*, 100–112.

Schiersmann, C. (2009). Selbstorganisation und Problemlösen als Eckpunkte einer allgemeinen Beratungstherapie. In H. Pühl (Hrsg.), *Handbuch Supervision und Organisationsentwicklung* (3. Aufl.). VS Verlag für Sozialwissenschaften.

Schmidbauer, W. (2009). Umsonst ist nicht vergebens – Konflikte an der Grenze von Ehrenamt und Profession. In H. Pühl (Hrsg.), *Handbuch Supervision und Organisationsentwicklung* (3. Aufl.). VS Verlag für Sozialwissenschaften.

Schrader, K. (2012). Politische Handlungsoptionen mit dem Ziel des Empowerments Drogengebrauchender Sexarbeiterinnen. *Standpunkt: Sozial 3*, Hamburger Forum für Soziale Arbeit und Gesundheit, S. 76–86.

Schreyögg, A. (2003). Die Differenzen zwischen Supervision und Coaching. *Organisationsberatung, Supervision, Coaching, 3*, 217–226.

Schreyögg, A. (2010). *Supervision. Ein integratives Modell* (5th ed.). VS Verlag für Sozialwissenschaften.

Semmer, N. K., Jacobshagen, N., Meier, L., Elfering, A., Kälin, W., & Tschan, F. (2013). Physische Beanspruchung durch illegitime Aufgaben. In Bundesanstalt für Arbeitsschutz und Medizin – G. Junghans, & M. Morschhäuser (Hrsg.), *Immer schneller, immer weiter: Physische Belastung bei Wissens- und Dienstleistungsarbeit* (S. 97–112). Springer.

Storch, M., & Krause, F. (2011). *Selbstmanagement – ressourcenorientiert. Grundlagen und Trainingsmaterial für die Arbeit mit dem Zürcher Ressourcen Modell* (4. Aufl.). Huber.

Thiersch, H. (1992). *Lebensweltorientierte Soziale Arbeit. Aufgaben der Praxis im sozialen Wandel*. Juventa.

von Foerster, H. (2003). Ethik und Kybernetik zweiter Ordnung. In P. Watzlawick & G. Nardone (Hrsg.), *Kurzzeittherapie und Wirklichkeit. Eine Einführung* (2. Aufl.; S. 71–90). Piper.

Vorheyer, C., & Nagel, U. (2011). Der habituelle Umgang mit den Paradoxien des professionellen Handelns: Soziale Arbeit in der Prostitutionsszene. In R. Becker-Lenz, S. Busse, G. Ehlert, & S. Müller (Hrsg.), *Professionelles Handeln in der Sozialen Arbeit* (S. 13–30). VS Verlag für Sozialwissenschaften.

Voß, G., Handrich, C., Koch-Falkenberg, C., & Weiß, C. (2013). Zeit- und Leistungsdruck in der Wahrnehmung supervisorischer Experten. In Bundesanstalt für Arbeitsschutz und Medizin – G. Junghans, & M. Morschhäuser (Hrsg.), *Immer schneller, immer weiter: Physische Belastung bei Wissens- und Dienstleistungsarbeit* (S. 63–96). Springer.

Winter, D. (2009). Arbeitsbedingungen in der Prostitution im Wandel von Zeit und Gesetz. In B. Kavemann & H. Rabe (Hrsg.), *Das Prostitutionsgesetz. Aktuelle Forschungsergebnisse, Umsetzung und Weiterentwicklung* (S. 221–230). Budrich.

Angela Siebold has insights into the development of the work of counselling centres for women in prostitution. She is a lecturer at various universities and a trainer at the Academy for Intercultural Action Competence.

Needs Analysis and Concept Development for a Counselling Centre in the Field of Prostitution for the City of Mannheim

Julia Wege and Martin Albert

1 Introduction

This article is based on a study on the situation of prostitution in the city of Mannheim (Wege, 2010). Within the framework of this study, a need for a psychosocial counselling centre for women in prostitution was identified. The study provided an important basis for the preparation and development of a basic concept for the establishment of a corresponding counselling centre. The preparation and planning period lasted almost three years. This long lead time made it possible to develop the concept in great detail and on a sound basis. It was of particular importance that the experience of other counselling centres was incorporated into the development of the concept. Discussions with nationwide experts and on-site visits in Dortmund, Berlin, Hamburg and Stuttgart also provided important information. Basic scientific research, a clear needs analysis and the establishment of a professional network are the most important building blocks for successful development in the concept phase. In the meantime, the Amalie Counselling Centre has been in

J. Wege (✉)
Mannheim, Germany
e-mail: julia.wege@rwu.de

M. Albert
Heidelberg, Germany
e-mail: martin.albert@srh.de

M. Albert, J. Wege (eds.), *Social Work and Prostitution*,
https://doi.org/10.1007/978-3-658-37761-8_13

existence since July 2013 and has essentially implemented all the central action steps and recommendations of the concept. Thus, the following explanations could provide important suggestions and hints for the establishment of corresponding counselling centres in other municipalities and regions.

2 Starting Position for the Concept Development

The city of Mannheim, as a metropolitan centre in the Rhine-Neckar metropolitan region and in particular due to its geographical location, has a specific significance and attraction for the area of prostitution. In the public context, both in the past and in the present, individual city districts such as Jungbusch and Neckarstadt-West have been associated with the topic in a variety of ways. These two districts show an increased need for social development and are characterised by a concentration of poverty, a high density of residents and an above-average proportion of migrants. In Mannheim, publicly perceptible prostitution takes place relatively close to the city centre, i.e. it is not exclusively located in outer urban fringe or remote commercial areas. Prostitution is clearly visible in the residential area of Neckarstadt-West and thus plays a role that should not be underestimated. However, it is important to point out that prostitution cannot be reduced to such local conditions. On the contrary, it can be clearly seen in Mannheim that this is an extremely complex and often imperceptible field of action. Assuming that there is no obligation for women who engage in prostitution to register and that after the introduction of the Prostitution Act the "compulsory examination" by the health authorities was abolished, it is understandable that hardly any reliable figures are available for Mannheim. Concrete data are definitely not possible due to the heterogeneity of the field of action. It should also not be underestimated that prostitution is often practiced only occasionally, for short periods of time and in different places or cities.

In this context, it is important to point out that Mannheim, as the second largest city in Baden-Württemberg, has specific dynamics for this area due to its social and economic structures. With regard to the extent of prostitution, a comparison with the scale of Stuttgart is helpful: according to findings of the Investigation Service Prostitution, 4000 women and 300 men engage in prostitution in Stuttgart (Gesundheitsamt Stuttgart, 2011). According to an evaluation of the relevant data and a conservative estimate by experts, the number for the Mannheim city area is probably at least 500 prostitutes (Wege, 2010, p. 64). If additional aspects are taken into account, such as prostitution by illegal migrants, prostitution in the drug sector, prostitution as a concealed street prostitution, prostitution in privately rented time-share apartments and the increasing prostitution mediation via the Internet, the number is likely to be much higher.

Throughout Germany, professional social counselling centres for prostitutes already exist in all large cities of comparable size. Even medium-sized cities such as Heilbronn have such institutions. In a report in 2001, the "Fachbereich Gesundheit" (Health Department) in Mannheim demanded that the deficits of the counselling services be compensated and that appropriate structures be established (Stadt Mannheim, 2009). Especially for migrants with illegal residence status, the counselling offer was "insufficient" or "not available at all". Human trafficking and the spread of sexually transmitted diseases represent a major problem area, especially in Mannheim. This concern remained unsuccessful for political reasons. Within the framework of a scientific study entitled "The need for social work action in the problem area of "prostitution" using the example of the city of Mannheim", it was determined after evaluating the available data and analysing extensive interviews with experts that there is a clear need for professional counselling structures in the area of prostitution in Mannheim and that a corresponding facility is urgently needed (Wege, 2010).

As a welfare organisation with a long tradition, the Diakonisches Werk has a wide range of advisory structures and experience in this field, both in Baden-Württemberg and nationwide. Due to the complexity of prostitution, the different local developments and the specific problems associated with it it seems logical to develop an in-depth basis for action before implementing such a project. In continuation of the study mentioned at the beginning, the Diakonisches Werk Mannheim commissioned the "Heidelberger Institut für Sozial- und Verhaltenswissenschaften" at the SRH Hochschule Heidelberg at the end of 2010 to develop a sustainable and custom-fit concept for corresponding counselling structures, which is based on the action steps shown in Fig. 1

On this basis, a concrete action concept will be developed for Mannheim with regard to target groups, focus of counselling, size of the counselling centre, staffing ratio and qualification of the employees. From this follows a proposal for a tempo-

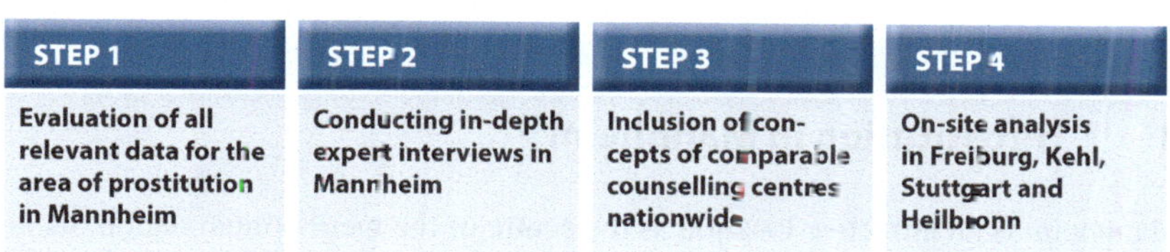

Fig. 1 Action steps of the scientific analysis. (Source: Own representation)

ral implementation of the concept and concluding recommendations for action. For the time being, the interview evaluation is based on the aforementioned preliminary investigation in the context of a master's thesis (Wege, 2010). The basis for the present counselling centre concept was the evaluation of the following additional expert interviews:

1. Social workers of the counselling centres in Stuttgart, Freiburg, Kehl and Heilbronn
2. Specialist counsellors of the Drogenverein Mannheim e. V.
3. Specialists of the "Fachbereich Gesundheit" and from the district Neckarstadt-West
4. Representatives from the local council
5. Representatives of public authorities

A total of 32 interviews were thus conducted with experts in the field, which thus form the basis for the creation of the concept. The practical implementation of the counselling concept requires a lot of patience and stamina. Due to the rather taboo and ambivalent handling of the topic, the associated objectives of the project are to be formulated very close to everyday life and pragmatically. Successful implementation can only be achieved if the responsible agency, in close cooperation with all offices, services and counselling centres – and here in particular the Mannheim city administration, the health department, the police, the welfare associations and the employment agency or job centre – assume joint responsibility for this. Such a counselling centre will only be able to work successfully in an association and within the framework of a trusting and network-oriented cooperation. It should be pointed out in particular that experience has shown that the sustainability of such a project can only be established after a period of three to five years.

This concept provides a sound basis for establishing successful and sustainable counselling structures. The goal must be to focus on the actual life needs of the women. This is also connected with the vision and the hope that in this way a new and social way of dealing with the topic of prostitution will succeed and a start for a public discourse within the city of Mannheim will be made possible.

3 Prostitution in Mannheim

Mannheim is an attractive location as the centre of the metropolitan region, its favourable transport infrastructure due to its proximity to Frankfurt Airport and the federal states of Hesse and Rhineland-Palatinate. Economic centres are also centres for the red light scene and for prostitutes, johns and pimps. Mannheim can look back on a long tradition of prostitution. In the Jungbusch district, prostitution used to cause quite a stir, especially due to the proximity of the harbour. In recent years, the scene has shifted to the inner city area and the Neckarstadt-West district, where most of it currently takes place. Drug prostitution in the 1990s was a problem area especially in the squares. In close cooperation with the police, the "Fachbereich Gesundheit" (Health Department), the Mannheim Drug Association and other authorities, it was possible to successfully counteract this. In 1976 the Regional Council of Karlsruhe (Regierungspräsidium Karlsruhe) declared the entire inner city of Mannheim a restricted area. According to the restricted area ordinance, prostitution is prohibited in this area, in an apartment, and even contact between a prostitute and her john within this district. If this restricted area ordinance is violated, the police authority threatens to ban the prostitute from the public area. If it is a case of procuring prostitution, the authorities must involve the drug association. Although these laws have clear guidelines and prostitution is strictly forbidden in certain areas of the city, the everyday reality in Mannheim is different. In the following, the forms of prostitution are first divided into three forms for a better overview (cf. Fig. 2). Afterwards, "visible prostitution" in Mannheim will be described in more detail, as it constitutes the largest and most important group and is of particular importance for the counselling centre.

Prostitution has shifted into a grey area in recent years, the reality of which is difficult for experts to assess. According to this, the current number of prostitutes is on average about 600 women. According to experts, the range is from approxi-

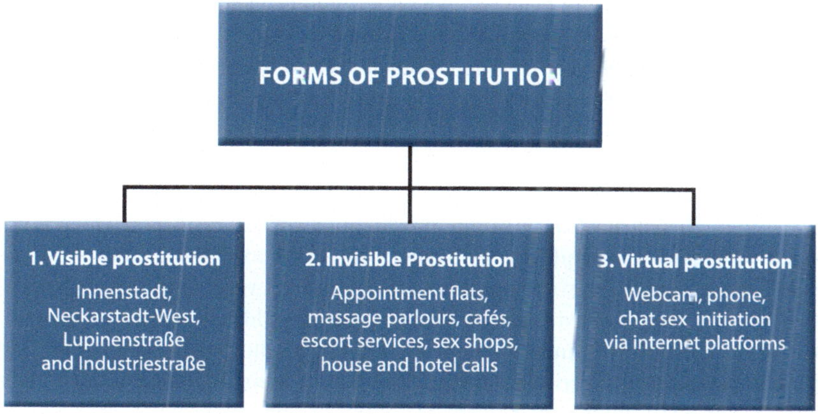

Fig. 2 Forms of prostitution in Mannheim. (Source: Own representation)

mately 500 to 1200 women, with the proportion of migrants being approximately 90% (Wege, 2010, p. 64). The women who engage in prostitution come mainly from the Balkan countries (Romania, Bulgaria, Poland, Ukraine, Belarus) or from Africa and have little or no knowledge of German. Sexual services are offered at different places in Mannheim: in the city centre, in Jungbusch, in Neckarstadt-West and in Industriestraße.

In the following, different places and their problems are presented in a differentiated way. In the city centre there are so-called red-light bars, where access is prohibited for those under 18 years of age. There, the women use the bar as a platform to get in contact with clients and also offer their services in the back rooms. In recent years, however, this so-called "bar prostitution" has declined because the police presence was too great and so the Lupinenstraße was preferred as a legal place for sexual services. Another problem in the inner city is drug prostitution. These women belong to a special group, as they prostitute themselves in order to be able to finance their own addiction or that of their partner. The street prostitution in the Industriestraße has also decreased in recent years, in the evening there are isolated women standing there or waiting in their illuminated cars. Although prostitution in Jungbusch has declined and shifted to Neckarstadt-West, there is talk there of Romanian and Bulgarian groups running an organised prostitution network. In Neckarstadt-West, which takes up the largest part, it is necessary to differentiate between two different areas: Lupinenstraße, also called Bordellstraße ("brothel street"), where about 120–140 women work and rent rooms there by the day, in which they also live for the most part. In the Neckarstadt-West district, there are additional expert indications of a form of prostitution organised by larger Eastern European "family clans" in which the family members act as a form of pimp. Here, prostitution is not seen as a profession, but as a family duty. These young women usually have no health knowledge, no access to medical care and offer their services for a very low price. The problems and repercussions that result are disturbing to local residents. The area from Mittelstraße to Industriesstraße is particularly problematic. Prostitution has become very widespread here, taking place during the day, at all hours, in playgrounds, next to schools, in cafes, private homes, etc. This means that prostitution is highly visible to residents, which is especially dangerous for children and young people. Parents' associations have already joined forces to do something about this, but so far without success. Also, the problem of young women from the Roma group should not be underestimated, who are of a certain size in Mannheim and are forced into prostitution. According to the experience of the criminal investigation department, it can be assumed that only a very small number of women are willing to prostitute themselves voluntarily. It must be taken into account that the motives for working in prostitution vary greatly and that the boundaries between prostitution and human trafficking can be blurred. Mannheim plays an important role in the area of human trafficking, both in Baden-Württemberg and in a nationwide comparison.

Although Mannheim directly borders Ludwigshafen, the red-light scene of both cities should not be mixed but considered separately, as they are two different cities that are not recognizably entangled with each other in this topic. According to the experts, there are about 300 prostitutes working in Ludwigshafen. So-called "appointment flats" are often found in the vicinity of large companies. The counselling centre "Solwodi" (Solidarity with Women in Distress) has been in existence since 2005 and was established with the financial support of "Aktion Mensch". The counselling centre is aimed at victims of human trafficking who have been abused for the purpose of sexual exploitation.

4 Consulting Concept for the City of Mannheim

Principles of a Mission Statement

Prostitution is an extremely complex field with very different manifestations and characteristics. Prostitution ranges from voluntary full-time practice to human trafficking and forced prostitution. With the introduction of the "Law on the Regulation of Legal Relationships of Prostitutes" (Gesetzes zur Regelung der Rechtsverhältnisse der Prostituierten – ProstG) on January 1, 2002, immorality was abolished and prostitution was legalised as a professional practice. Prostitutes can register and conclude employment contracts under this professional title and are thus subject to compulsory social insurance. In addition, there is the possibility of legal action in the case of outstanding payment for services. This was connected with the hope that prostitution would receive social and professional legalisation and social recognition. In reality, however, prostitution continues to move in a dichotomy of stigmatisation and exclusion. For many reasons, social recognition of prostitution as a profession cannot be ascertained and, depending on the perspective, is also doubted. As a rule, the work of prostitutes still takes place in secrecy or anonymity. Socially, the topic of prostitution continues to be taboo and forms the basis for clichéd images and voyeuristic assessments (Albert & Wege, 2011). The discussion in public and especially in politics is morally ambivalent. On the one hand, it contradicts social values of love, partnership, sexuality and fidelity. On the other hand, prostitution is a social reality of great magnitude. This tension, especially from which perspective prostitution is viewed and evaluated, is of a fundamental nature and raises more questions than answers: Between personal value attitudes and found practice and reality, a tense ambivalence remains, which will probably not be resolved.

The manifestations of prostitution are manifold and influence the moral evaluation. On the one hand, it ranges from the attitude that people make themselves available for this "service" in return for payment and voluntarily, and that this represents a self-chosen perspective on life. However, it is undoubted that large areas of prostitution are subject to a variety of coercive mechanisms and social conditions in which experiences of violence and dependence, ignorance of one's own rights, criminalisation and pressure due to competition prevail (BMFSFJ, 2009). In this context, it is not to be underestimated that a large part of the mostly younger female prostitutes are migrants, especially from Eastern European countries. Poverty and lack of job prospects are probably the main motives for engaging in prostitution. The desire to return and to exercise a legal profession as well as to build up a positive perspective on life is a decisive factor for this target group.

Taking into account the ambivalence of the entire field of action, it is urgently necessary for the counselling centre to work on the basis of a mission statement. The focus of ethical action is to respect the rights and dignity of women and to be oriented towards their life needs. An important prerequisite for this is openness, appreciation, acceptance and an avoidance of moral judgements. The following principles could provide initial orientation for the activities of the counselling centre:

1. Women have the right to exercise their activity in a self-determined and free manner.
2. The dignity and rights of persons engaged in prostitution shall be respected and upheld.
3. Any act of prostitution carried out under psychological and physical coercion must be rejected.
4. Counselling is available to all women, regardless of the form of prostitution they engage in.
5. Social counselling and care is provided without restrictions. It is comprehensive, holistic and oriented to the needs of the reality of life of the women.
6. Women have the right to be informed about the psychological and health effects of prostitution.
7. The counselling aims to show new perspectives outside of prostitution and offers concrete exit assistance.
8. The counselling centre or the responsible body actively campaigns for social recognition with regard to the specific life situation of prostitutes and thus opposes public exclusion, stigmatisation and criminalisation.

These principles provide both the provider and the supporting network with a basis for appreciative cooperation. Especially for the employees working in this ambivalent field, such guidelines are indispensable and offer security for the focal points of activity. The design of the framework conditions for the counselling centre should be explicitly oriented towards these principles and value standards and offers the basis for a further development of the mission statement.

Target Groups

The activities of the counselling centre are initially aimed at all women who engage in prostitution. In a broader sense, this target group is understood to include only women for the time being, i.e. not men and transsexuals. A gender-specific separation in the context of prostitution counselling exists in many counselling centres in comparable cities. There are no data and information available for Mannheim on the special situation of male prostitutes.

Against this background, the counselling centre is aimed at women,

- who engage in prostitution on a professional and voluntary basis (e.g. in brothel-like establishments, so-called "Laufhäuser" (running houses), appointment flats, etc.),
- with a migrant background who engage in prostitution partly because of the financial support of their families in their home country,
- who engage in prostitution as a "casual job" in addition to their legal work,
- who work on the street or in mobile homes with changing locations,
- who use prostitution to finance their drug use,
- who have committed borderline prostitution offences and are temporarily detained,
- who work as prostitutes on an occasional or regular basis via the internet,
- operating in the broader area of prostitution (e.g. upmarket "sauna clubs", escort services, sex shops, animation bars, etc.),
- with illegal legal status or engaged in illegal prostitution,
- who have left the field of prostitution and need appropriate psychosocial care,
- who are affected by human trafficking and forced prostitution,
- and young girls of minor age who engage in prostitution voluntarily or under coercion.

As already pointed out, prostitution is an area that is subject to constant change. In this sense, there are only to a limited extent clearly defined target groups. Places of residence and workplaces are constantly changing and are not subject to clear structures. The counselling centre should be open to all the target groups described and will be confronted with new forms of prostitution in the future, especially promoted by the new media and the Internet.

Task

The tasks of the counselling centre should be characterised by openness to all life questions and needs of the counsellors. Prostitution is subject to a high degree of taboo, so it is to be expected that there will be little willingness to openly admit to this activity at the beginning. There is also no clear profile of the questions the women come to the counselling centre with. In this context, it must be pointed out that a large proportion of the women come from Eastern Bloc countries, are comparatively young and may only be in the area for a short period of time. The "market" and its mechanisms require the women to change locations involuntarily and constantly. Added to this is the fact that the workplace and the "sleeping place" are often identical. Against this background, social relationships can hardly be built up and require a specific approach with a medium-term orientation for counselling activities. Experience shows that practical assistance with everyday problems (e.g. information on doctors, health insurance, visits to the authorities) is the first step towards counselling. At first, the women expect quick and concrete assistance that does not have anything to do with the actual activity in the prostitution sector. This kind of contact can be the beginning of a possible long-term cooperation, which can lead to an opening for all areas of coping with life. The counselling activities should include the following areas of responsibility:

1. General social counselling on all issues related to prostitution activities
   - Information about the legal framework
   - Education and counselling on general health and hygiene issues and implications.
   - Help with financial problems and debt counselling
   - Assistance in finding accommodation or in case of housing problems
   - Information on health insurance coverage, social security and pension insurance
   - Advice and assistance in court proceedings
   - Provision of instrumental assistance (e.g. condoms, hygiene articles, clothing, etc.)

- Orientation, assistance and mediation with regard to specialist counselling centres, authorities and doctors
- Clarification of issues relating to the law on aliens

2. Psychosocial counselling
   - Comprehensive intervention for severe life crises
   - Assistance in the event of violence
   - Counselling and support during imprisonment
   - Education about the psychological and physical effects of prostitution
   - Counselling for loss of social contacts, isolation and emotional dependencies
   - Assistance with partnership and family problems
   - Counselling and support during pregnancies and related follow-up problems or questions
   - Anonymous telephone counselling for all life questions
3. Target group oriented consulting
   - Goal-oriented counselling for women with a short local stay (approx. 10 weeks)
   - Counselling of migrants and information on return assistance to the home countries
   - Targeted care for women with children
   - Care of underage girls engaged in prostitution under duress or voluntarily
   - Initiation and support of self-help groups for prostitutes
   - Aftercare and accompaniment of former prostitutes
4. Exit counselling and targeted case management
   - Comprehensive orientation advice
   - Information about the labour market and opportunities to enter new fields of work (e.g. care sector)
   - Job application support and mediation of professional qualifications
   - Network-oriented referral to specialist counselling centres (e.g. addiction counselling centres) and authorities (e.g. employment agency, job centre)
   - Establishment of contacts to institutions in the home countries for support upon return
5. Networking
   - Initiation and implementation of a "Round Table on Prostitution"
   - Close cooperation and collaboration with the municipal "Fachbereich Gesundheit" (Health Department)
   - Informing the specialist authorities and the police about the activities of the advice centre
   - Targeted cooperation with all municipal offices and authorities

   - If necessary, exchange and cooperation with operators of brothel-like establishments
6. Public relations and lobbying
   - Creation of advertising materials for the counselling centre
   - Recruitment and involvement of sponsors for special project work
   - Targeted press work with regard to de-stigmatising prostitution
   - Recruitment and involvement of volunteers in counselling activities
   - Prevention work, e.g. in youth facilities and on the Internet
   - Internet presence in different languages
   - Representation and representation of the advice centre in political bodies
   - Cooperation with nationwide counselling centres for prostitutes

Of particular importance to the counselling centre is that the work should have a clear "going structure" – as opposed to a so-called "coming structure". Women who work in the stigmatising field of prostitution will probably be reluctant to visit the counselling centre at first. The women concerned need an open offer of help, which should allow access on the spot without conditions. Thus the objective of this form of outreach work is geared towards holism and closeness to everyday life. For professional counsellors, this means that they should also offer their services in the local area of prostitution during evening hours until midnight. The entry into the work of the counselling centres can take place through the distribution of information brochures, hygiene articles and condoms, etc. Appointment flats, bars and running houses are to be visited directly. In this context, it is important to point out that such on-site visits should generally only be made in pairs.

Project Work

The work of the counselling centre should additionally include high-profile projects. It is only through innovative ideas and projects that the work becomes known and also enables sponsors to become sustainably involved in support structures. Throughout Germany, the corresponding specialist counselling centres have extremely interesting locally oriented projects. It should be examined which of these projects can be transferred to the local conditions of Mannheim. In this context, the location of the counselling centre will not be insignificant. In the following, a number of project ideas are proposed that are feasible for Mannheim.

- Creation of guidelines regarding authorities, emergencies, legal regulations
- Offer of a medical consultation by doctors and physiotherapists within the counselling centre and establishment of a regular women's café

- Initiation of cultural projects in cooperation with theatre institutions and other cultural institutions
- support measures involving women who are still active in prostitution
- Targeted use of voluntary helpers (e.g. as "pilots", accompaniment to visits to the authorities)
- Offer of a clothing store or laundry to be washed
- Promotional campaigns at Christmas or on other occasions (e.g. giving every woman a red rose on Valentine's Day with an address and information about the counselling centre).

Experience has shown that the willingness to become involved as a sponsor for the objectives of the counselling centre and the concerns of the different target groups can be considered high. Mannheim offers a great potential for development due to its economic, social and cultural conditions and with the appropriate targeted support.

Cooperation with the "Health Department"

Of central importance is the cooperation or the integration of the "health department" in the activities of the counselling centre. After evaluating the expert interviews of all counselling centres in Baden-Württemberg, it can be clearly seen that the access structures from this side into the milieu are particularly important. At the counselling centre in Stuttgart, for example, the social workers are directly employed by the Department of Health, but work at the local counselling centre, which is supported by Caritas.

Before the introduction of the Prostitution Act (Prostitutionsgesetzes), the social service of the "health department" in Mannheim was an essential access route. In the context of the revision of the legal provisions, the lack of structures was pointed out especially by the health department. Women's access to these counselling services is now only limited and rudimentary and is thus only used when necessary. It is not insignificant that in this context is based on activity on the part of participants and women only visit the "health department" when needed. Although questions about health protection, hygiene and the risk of STIs are of great importance, women with a migration background and a short length of stay hardly make use of such traditional counselling concepts. A special target group may be women with unwanted pregnancies. According to experts, abortions are very often performed and raise a number of problems. The integration of the "health department" into the concept of the counselling centre can take place in a variety of ways. Special offers in the counselling centre up to joint visits on site in brothel-like establishments or of appointment flats are conceivable.

Staffing, Location and Size of the Counselling Centre

The recommendations on the staffing of the counselling centre include statistical data such as the population of Mannheim, estimated figures from experts and comparative data from other counselling centres. It is also taken into account that the personnel resources in other cities are financed by targeted project funds with a corresponding objective. In this context, special reference should be made to funding in relation to exit counselling or psychosocial support for women in the context of forced prostitution. Taking all aspects into account, we consider a staffing ratio of at least two full positions of specialist staff and half a position of administrative staff to be urgently required. It is urgently not advisable to staff the counselling centre with only one position. Experience has shown that extensive administrative and clerical work is associated with such advisory work (e.g. writing letters to the authorities, financial planning, maintaining office space, preparing reports, etc.), from which the expert advisers should be relieved. For this reason, a 50% position for an administrative specialist to work exclusively on this administrative area is strongly recommended. With reference to the filling of two social worker posts, we tend to consider a distinction within the framework of the content of activities to be sensible when considering professional qualifications. In this respect, the following focal points of work with corresponding qualifications would be recommendable:

1. Staff position: Focus on psychosocial counselling
   - Individual case assistance and case management
   - Street work
   - Organisation and implementation of an open and low-threshold meeting place
   - Social therapeutic interviewing
   - Qualification and corresponding professional experience in therapy and counselling methods, trauma processing and crisis intervention
   - Degree: Master of social work with focus on counselling
2. Staff position: Focus on network and public relations work
   - Case management and exit counselling
   - Public relations, media and lobbying
   - Initiation and moderation of the "Round Table on Prostitution"

- Initiation of sponsoring activities and support of voluntary projects
- Financial planning and representation in specialist policy bodies
- Qualification and relevant professional experience in the field of moderation, community and network work
- Degree: Master of Social Work with a focus on social management

A successful staffing contributes significantly to the development of the counselling centre, to its successful success and to its general standing in the social sector of Mannheim and requires a careful selection process. The work requires a high degree of self-reflection and professional external perception as well as the ability for conceptual development. Previous professional experience in the counselling of prostitutes is not absolutely necessary, especially as this is a narrowly defined professional area throughout Germany. What is more important is a high level of professional commitment in the initial phase, the ability to establish contact with the women or to gain their trust, and to secure the long-term future of the counselling centre through targeted financial and project support. Strong public interest in the activity (e.g. local media enquiries) is to be expected, in this respect the counsellor should be capable of public and media relations. The job profile requires in particular: the ability to initiate networks, broad-based competencies in individual case assistance, skills that enable complex organisational management, a high level of psychological resilience and flexibility in working hours. A stable personality and a high degree of self-reflection are also essential. A commitment to ongoing training and supervision is essential. The required professional qualifications and the expected stresses of the job should also be reflected in the pay. Even though the number of staff positions is comparatively manageable, we consider a clear management structure necessary due to the special nature of the counselling centre.

With regard to the location and the spatial size, the local proximity, the form and the objective should basically be taken into account. Each of the nationwide existing specialised counselling centres has a specific integration into the social organisational structure and a public presence of the respective municipality. A spatial proximity to one of the focal points of prostitution in Mannheim or in Neckarstadt-West is recommended. Here, the existing structures of the Diakonisches Werk Mannheim could be used. The work in the area of prostitution moves between latent taboo, subtle criminalisation and social defence, whereby the actual interests of the women concerned are given little consideration. Some of the women from this milieu have experienced violence and trauma and need a protected framework. The women's wish for "anonymous" counselling must be respected and thus also has a decisive influence on the choice of location. After evaluating the expert interviews, it was frequently pointed out that the proximity, but also the openness of the counselling structure (e.g. in the form of an open café-meeting place) should be taken into account. The location has a major impact on the success of the counselling centre.

In order to be able to carry out high-quality work in the counselling centre, premises are required that have sufficient space. In order to meet these requirements, the counselling centre should have at least 130m^2 of usable space. Preference would be given to a self-contained flat or shop with at least five rooms. These would be divided as follows:

- An open lounge area of approximately 30 m^2
- An office space for the administration of approximately 10 m^2
- Two consulting rooms of approximately 15 m^2 each
- A room for group activities of approximately 20 m^2
- A room for external counselling services (medical counselling, legal counselling, etc.) of approximately 10 m^2

The open lounge area should have an appropriate kitchenette and offer the possibility for a flexible first contact. A professional room for medical counselling, as is the case in Stuttgart, for example, would be desirable. Such an offer would enormously increase the access to the counselling centre or the interest in taking advantage of the offer at all. It would also be advisable to provide additional cellar or storage space (e.g. for clothing donations, etc.). The counselling centre should be located in Neckarstadt-West and, if possible, near the corresponding brothel street (Lupinenstraße) in order to create a public and noticeable presence in the district.

Round Table on Prostitution

Due to the complexity of the field of action, we consider the establishment of a "Round Table on Prostitution" for Mannheim absolutely necessary. The current situation in Mannheim is characterized by a lack of information and a low level of coordination between the agencies involved. The round table should deal with all questions and problem areas concerning prostitution in the Mannheim area and the neighbouring Rhine-Neckar region. One basis for this is the implementation and design of the current Prostitution Act in order to discuss, among other things, low-threshold exit assistance in a larger context and to mutually clarify appropriate measures. The common goal should be an overarching concept for Mannheim in which the cooperation and coordination of those involved is defined. The round table should consist of the following institutions:

- Representatives of municipal authorities
- Representatives of the local council and political parties
- Department of health
- Employment agency
- Police and representatives of relevant departments
- Charities and churches
- Independent organisations, self-help groups and initiatives
- Additional social institutions from the Rhine-Neckar region

At the beginning, the advisory body can act as a provider of ideas and initiator of this round table, although in the medium term this would require a political decision by the municipal council. Only on a broad local political basis would a round table have the legitimacy not only to formulate more far-reaching recommendations but also to implement them in practice. Corresponding meetings would have to take place at least every quarter. In order to increase the weight of the "Round Table on Prostitution", it should have the right to formulate corresponding questions and test assignments, such as for the licensing of brothels or the residence regulations for victims of human trafficking. The Round Table should be based on the general agreement that prostitution is not an immoral occupation and that the living and working situation of prostitutes should be improved. However, an essential goal remains to offer exit assistance and to counteract all forms of exploitation and forced prostitution. Only on the basis of mutual exchange, overarching cooperation and the professional examination of specific problem areas can it be possible to define well-founded recommendations for action and to enable their implementation. This is supported by the common will to create a balance between the interests of women who engage in prostitution and the general population. This also makes an important contribution to social peace in the local context.

5 Concrete Recommendations for Action

Considering all aspects that are relevant for the area of prostitution in Mannheim, we consider the establishment of professional counselling structures to be urgently necessary. Prostitution in Mannheim has an enormous relevance and is connected with high social effects. Especially the tabooing of the topic in the public context and the "invisibility" of the related problem areas must be an incentive to establish such a counselling centre as soon as possible. The creation of professional support services for women working in the field of prostitution is an important task for the social landscape and the public, which all institutions involved must deal with openly and without reservations. In order to establish the counselling centre professionally, we recommend the following action steps:

1. The future counselling centre needs a coherent and meaningful concept, which must be geared to the everyday and living needs of women and target groups working in the field of prostitution.
2. It is recommended to present the counselling centre to be established professionally and offensively to the public. This requires an innovative "communication and advertising concept" with logo and design to ensure sustainability in the city of Mannheim.
3. For a longer-term identification, a self-developed concept for the name of the counselling centre is of high importance. When choosing the name, the wishes and interests of the women should be given special importance.
4. The necessity of the counselling centre requires a clear socio-political expression of will from the public side. The basic financing must be guaranteed by public subsidies and here in particular by the city of Mannheim.
5. We recommend setting up the counselling centre at a separate location in order to enable low-threshold and trusting access. The establishment in Neckarstadt-West seems to be logical and reasonable due to the problematic situation, the local access possibilities and the structural conditions.
6. The counselling centre needs sufficient material, instrumental and personnel resources for the successful implementation of its goals and tasks. An apartment or shop size of approx. 130 m2 and the employment of at least two social worker positions as professional staff and half an administrative staff is a fundamental prerequisite for the start and further development.
7. We recommend the use of mother-tongue counselling services in order to meet the everyday needs of the different migrant groups.
8. We consider the cooperation with the municipal "Department of Health" to be important for the success of the counselling centre. Special counselling offers should be developed jointly and cooperatively in this sense.
9. We recommend creating a separate and tailored concept for the involvement of volunteers. With regard to the acquisition of sponsors, we see great development opportunities for target group-oriented projects.
10. In order to accompany and support the counselling centre in the process of its further development, we recommend the establishment of an advisory board, in which public figures from Mannheim should also be represented.
11. For Mannheim, the establishment of a "Round Table on Prostitution" seems to be urgently necessary in order to ensure mutual information, coordination and

cooperation of all authorities, institutions and social facilities. The future counselling centre could be a first initiator and networking point for this.

It is urgently necessary to subject the project to a scientific external interim evaluation after three years. According to expert assessment, an evaluation of the sustainability and "success" of the project can only be made in a time frame of five years.

References

Albert, M., & Wege, J. (2011). Soziale Arbeit und Prostitution – Handlungsbedarf und Entwicklungsmöglichkeiten in einem tabuisierten Berufsfeld. *Soziale Arbeit, 1*, 8–18.

Bundesministerium für Familie, Senioren, Frauen und Jugend (BMFSFJ). (Hrsg.). (2009). *Regulierung von Prostitution und Prostitutionsstätten – ein gangbarer Weg zur Verbesserung der Situation der Prostituierten und zur nachhaltigen Bekämpfung des Menschenhandels*? Berlin.

Gesundheitsamt Stuttgart. (Hrsg.). (2011). *Sozialplanung für Prostitution*. http://www.stuttgart.de/item/show/303496. Zugegriffen 8. Feb. 2015.

Stadt Mannheim. (2009). *Informationsvorlage. Prävention sexuell übertragbarer Erkrankungen bei Prostituierten – Situation in Mannheim und Auswirkungen des neuen Infektionsschutzgesetzes (IfSG)*. Aktenzeichen 625/2001. Mannheim 2001. http://web.mannheim.de/webkosima/webkosima-_vorlagen/666_2001.pdf. Zugegriffen 1. Okt. 2009.

Wege, J. (2010). *Der sozialarbeiterische Handlungsbedarf im Problembereich "Prostitution" am Beispiel der Stadt Mannheim*. Unveröffentlichte Master-Thesis, SRH Hochschule Heidelberg.

Julia Wege Social Worker (M.A.), is head of the counselling centre "Amalie" – for women in prostitution in Mannheim. Since 2009 she has been a lecturer at the SRH Hochschule Heidelberg for individual case assistance, community work and trauma work. Her research and publications focus on prostitution, homelessness and community work.

Martin Albert Social Worker (M.A.), is head of the social work program at the SRH Hochschule Heidelberg. His focus is on social work science, professionalization, migration and community work. He has conducted various scientific studies in the fields of street work, district development and prostitution

The manufacturer's authorised representative in the EU is Springer Nature Customer Service Centre GmbH, Europaplatz 3, 69115 Heidelberg, Germany. If you have any concerns regarding our products, please contact ProductSafety@springernature.com

Printed and bound by CPI Group (UK) Ltd, Croydon, CR0 4YY
30/09/2026
02225448-0001